Communications
in Computer and Information Science 1541

More information about this series at https://link.springer.com/bookseries/7899

Bernabé Dorronsoro · Farouk Yalaoui ·
El-Ghazali Talbi · Grégoire Danoy (Eds.)

Metaheuristics and Nature Inspired Computing

8th International Conference, META 2021
Marrakech, Morocco, October 27–30, 2021
Proceedings

 Springer

Editors
Bernabé Dorronsoro (iD)
University of Cadiz
Cadiz, Spain

Farouk Yalaoui (iD)
Université de Technologie de Troyes
Troyes, France

El-Ghazali Talbi (iD)
University of Lille/Inria
Lille, France

Grégoire Danoy (iD)
University of Luxembourg
Esch-sur-Alzette, Luxembourg

ISSN 1865-0929   ISSN 1865-0937 (electronic)
Communications in Computer and Information Science
ISBN 978-3-030-94215-1   ISBN 978-3-030-94216-8 (eBook)
https://doi.org/10.1007/978-3-030-94216-8

This Springer imprint is published by the registered company Springer Nature Switzerland AG
The registered company address is: Gewerbestrasse 11, 6330 Cham, Switzerland

Preface

This book compiles the best papers submitted to the Eighth International Conference on Metaheuristics and Nature Inspired Computing (META 2021). META 2021 took place in Marrakech, Morocco, from October 27 to 30, in a hybrid onsite/online mode. The main objective of META 2021 was building an atmosphere for the exchange of knowledge on the topic of evolutionary computation and nature inspired computing, where researchers in the field from all over the world could present their recent works and discuss them with other colleagues. META is a biannual conference that has been organized in Morocco since 2014, with the objective of establishing bridges between developing countries in Africa and developed countries from all over the world. Particularly, META aims at attracting influential researchers worldwide in the fields of metaheuristics and nature inspired computing for complex problems optimization.

Three categories of papers were considered in META 2021, namely work-in-progress and position papers, high impact journal publications (in the shape of an extended abstract), or regular papers with novel contents and important contributions. The conference received a total of 53 papers, which were evaluated using a blind review process.

Thirty five papers were presented during the META 2021 edition, which were arranged into five sessions, including one large session with video presentations and four sessions with both onsite and online presentations and audiences. The sessions covered topics such as the design of novel optimization tools, the synergies between optimization methods and learning techniques, and applications of such tools to real-world complex problems.

A selection of the best 16 regular papers is published in this book, representing a 30.19% acceptance rate for all submitted papers. The papers were selected according to the scores received in the blind review process. We hope that you enjoy reading them and find inspiration for future research.

October 2021

Bernabé Dorronsoro
Farouk Yalaoui
El-Ghazali Talbi
Grégoire Danoy

Organization

Conference Chairs

Bernabé Dorronsoro   University of Cadiz, Spain
Farouk Yalaoui   Université de Technologie de Troyes, France

Program Chairs

Amir Nakib   Université Paris-Est Créteil, France
Grégoire Danoy   University of Luxembourg, Luxembourg

Steering Committee Chairs

El-Ghazali Talbi   University of Lille and Inria, France
Rachid Ellaia   EMI, Morocco
Khaled Mellouli   Institut Supériour de Gestion de Tunis, Tunisia

Publicity Committee

Grégoire Danoy   University of Luxembourg, Luxembourg
Hatem Masri   University of Bahrain, Bahrain
Sergio Nesmachnow   Universidad de la República, Uruguay
Méziane Aider   Université des Sciences et de la Technologie
                                    Houari Boumediene, Algeria

Program Committee

M. Aider   Université des Sciences et de la Technologie
                                    Houari Boumediene, Algeria
E. Alba   University of Málaga, Spain
L. Amodeo   Université de Technologie de Troyes, France
R. Battiti   University of Trento, Italy
J. Blazewicz   Poznan University of Technology, Poland
C. Blum   Spanish National Research Council, Spain
P. Bouvry   University of Luxembourg, Luxembourg
T. Crainic   Université du Québec à Montréal, Canada
F. D'Andreagiovanni   Sorbonne University, France
K. Deb   Michigan State University, USA
J. del Ser   TECNALIA, Spain

K. Doerner	University of Vienna, Austria
M. Dorigo	Université Libre de Bruxelles, Belgium
A. El Afia	ENSIAS, Mohamed V University, Morocco
R. Ellaia	EMI, Morocco
T. Macias-Escobar	Instituto Tecnológico de Ciudad Madero, Mexico
P. Festa	University of Napoli FEDERICO II, Italy
J. Figueira	Technical University of Lisbon, Portugal
E. Fraga	University College London, UK
J. M. García-Nieto	University of Málaga, Spain
M. Gendreau	École Polytechnique de Montréal, Canada
F. Hanafi	Université Polytechnique Hauts-de-France, France
J.-K. Hao	University of Angers, France
R. Hartl	University of Vienna, Austria
L. Idoumghar	University of Haute-Alsace, France
I. Kucukkoc	Balikesir University, Turkey
P. Korosec	Jozef Stefan Institute, Slovenia
S. Krichen	Université de Tunis, Tunisia
R. Magán-Carrión	University of Granada, Spain
S. Martins	Universidade Federal Fluminense, Brazil
R. Massobrio	Universidad de la República, Uruguay
N. Melab	University of Lille, France
A. J. Nebro	University of Málaga, Spain
S. Nesmachnow	Universidad de la República, Uruguay
J. F. Oliveira	INESCTEC, Portugal
I. H. Osman	Lebanese University, Lebanon
G. Papa	Jozef Stefan Institute, Slovenia
M. Pavone	University of Catania, Italy
D. Precioso	University of Cadiz, Spain
J. Puchinger	Université de Lorraine, France
G. Raidl	University of Vienna, Austria
H. Ramalhinho	Universitat Pompeu Fabra, Spain
M. Resende	University of Washington, USA
C. Ribeiro	Universidade Federal Fluminense, Brazil
A. Salhi	University of Essex, UK
A. Santiago	Polytechnic University of Altamira, Mexico
F. Saubion	Université d'Angers, France
A. Sbihi	Brest Business School, France
O. Schütze	CINVESTAV, Mexico
P. Siarry	Université Paris-Est Créteil, France
T. Stutzle	Université Libre de Bruxelles, Belgium
P. Toth	University of Bologna, Italy

K. C. Tan	National University of Singapore, Singapore
A. Viana	Instituto Superior de Engenharia do Porto, Portugal
F. Yalaoui	Université de Technologie de Troyes, France
I. Yusuf	University of Malaya, Malaysia
N. Zufferey	Université de Geneve, Switzerland

Contents

Combinatorial Optimization

A Large Neighborhood Search for a Cooperative Optimization Approach to Distribute Service Points in Mobility Applications

Thomas Jatschka[1]([✉]), Tobias Rodemann[2], and Günther R. Raidl[1]

[1] Institute of Logic and Computation, TU Wien, Vienna, Austria
{tjatschk,raidl}@ac.tuwien.ac.at
[2] Honda Research Institute Europe, Offenbach, Germany
tobias.rodemann@honda-ri.de

Abstract. We present a large neighborhood search (LNS) as optimization core for a *cooperative optimization approach* (COA) to optimize locations of service points for mobility applications. COA is an iterative interactive algorithm in which potential customers can express preferences during the optimization. A machine learning component processes the feedback obtained from the customers. The learned information is then used in an optimization component to generate an optimized solution. The LNS replaces a mixed integer linear program (MILP) that has been used as optimization core so far. A particular challenge for developing the LNS is that a fast way for evaluating the non-trivial objective function for candidate solutions is needed. To this end, we propose an *evaluation graph*, making an efficient incremental calculation of the objective value of a modified solution possible. We evaluate the LNS on artificial instances as well as instances derived from real-world data and compare its performance to the previously developed MILP. Results show that the LNS as optimization core scales significantly better to larger instances while still being able to obtain solutions close to optimality.

1 Introduction

The traditional approach for solving service point placement problems, such as distributing charging stations for electric vehicles or vehicle sharing stations in a geographic area, essentially is to first estimate the demand that may be fulfilled at potential locations and then to select actual locations either manually or by some computational optimization. However, estimating the customer demand that may be fulfilled by certain stations is an intricate task in which erroneous assumptions may result in heavy economic losses for the service point provider. Also, estimating demand upfront requires specific data which can be challenging and/or expensive to collect. As an alternative approach, in [1] we introduced a

Thomas Jatschka acknowledges the financial support from Honda Research Institute Europe.

B. Dorronsoro et al. (Eds.): META 2021, CCIS 1541, pp. 3–17, 2022.
https://doi.org/10.1007/978-3-030-94216-8_1

cooperative optimization approach (COA) for optimizing the locations of service points in mobility applications. In contrast to the traditional approach, COA is an iterative interactive algorithm that solves the demand data acquisition and optimization in a single process by allowing customers to express their preferences intertwined with the optimization. A machine learning component processes the feedback obtained from the customers and provides a surrogate objective function. This surrogate objective is then used in an optimization component to generate an optimized solution. This solution is then a basis for further interaction with the users to obtain more relevant knowledge, and the whole process is repeated until some stopping criterion is met. So far, COA uses a mixed integer linear program (MILP) in the optimization core for determining solutions [2] or, in a former version [3], basic metaheuristic approaches that treated the problem as black box model and hence do not make significant use of structural properties of the problem. For an exact optimization core, the generated solutions are optimal w.r.t. to the so far known information derived from the customer feedback. However, this optimality comes at the cost of large computation times, especially for large-scale instances with thousands of customers and hundreds of potential service point locations. In contrast, a heuristic optimization core may feature better scalability towards larger instances. To this end we present here a large neighborhood search (LNS) that can reduce computation times by orders of magnitudes with only small losses in final solution quality. Due to the nature of the non-trivial objective function of our service point distribution problem, an efficient way for evaluating said objective is necessary to make this speedup possible. Therefore, our LNS features a data structure, referred to as *evaluation graph* for modeling the evaluation of solutions. We show how the evaluation graph can be used to efficiently keep track of small changes in the solution, such as opening or closing a service point. Based on this evaluation graph, the LNS is able to quickly repair partially destroyed solutions in a promising heuristic way. We evaluate the LNS on artificial instances as well as instances derived from real-world data and compare its performance to the previously developed MILP-based approach.

In the next section we review related work. Section 3 formally defines the General Service Point Distribution Problem (GSPDP), as it is referred to, while an overview on the COA framework is given in Sect. 4. Our main contribution, the LNS with its evaluation graph, is presented in Sect. 5. Section 6 explains the benchmark scenarios, and Sect. 7 discusses experimental results. Finally, Sect. 8 concludes this article and gives an outlook on future work.

2   Related Work

The basic concept of COA was presented in [1]. In interactive optimization algorithms, such as COA, humans are used to (partially) evaluate the quality of solutions and to guide the optimization process. For a survey on interactive optimization, see [4]. Interactive algorithms are often combined with surrogate-based approaches [5,6], in which a machine learning model is trained to evaluate intermediate solutions approximately in order to reduce user interactions and to avoid user fatigue [7]. In contrast to COA, most approaches from literature only allow a single user to interact with the algorithm, e.g., [8,9]. Hence, in [10] COA's surrogate function is based on a matrix factorization model [11], a popular collaborative filtering technique [12] in which unknown ratings of items are derived from users with similar preferences.

In [3] two heuristic black box optimization approaches were suggested for COA to generate new candidate solutions w.r.t. to the current surrogate model: a variable neighborhood search as well as a population-based iterated greedy approach. In [2] COA was substantially extended to also be applicable in use cases where the satisfaction of demands relies on the existence of two or more suitably located service stations, such as car and bike sharing systems.

More generally, there exists a vast amount of literature regarding the location planning of service points for mobility applications, see, e.g., [13] for electric vehicle charging stations or [14] for stations of a bike sharing system. However, to the best of our knowledge no further work on interactive optimization approaches for location planning in mobility applications exists.

3   The General Service Point Distribution Problem

In this section we give a formal description of the *Generalized Service Point Distribution Problem* (GSPDP) introduced in [2], which is the problem to be solved at the core of COA and for which we will then propose the LNS. Service points may be set up at a subset of locations $V = \{1, \ldots, n\}$. Establishing a service point at a location $v \in V$ is associated with costs $z_v^{\text{fix}} \geq 0$ and the total setup costs of all stations must not exceed a maximum budget $B > 0$. Additionally, the expected costs for maintaining this service point over a defined time are $z_v^{\text{var}} \geq 0$. Given a set of users U, each user $u \in U$ has a certain set of *use cases* C_u, such as going to work, visiting a recreational facility, or going shopping.

Each user's use case $c \in C_u$ is associated with a demand $D_{u,c} > 0$ expressing how often the use case is expected to happen within some defined time period. The demand of each use case may possibly be satisfied by subsets of service points to different degrees, depending on the concrete application and the customer's preferences. Hence, we associate each use case c of a user u with a set of *Service Point Requirements* (SPR) $R_{u,c}$ with which a user can express the dependency on multiple service points to fulfill the needs of the use case. For example, for the use case of visiting a fitness center using a bike sharing system, one SPR may represent the need of a rental station close to home or work and a second SPR a rental station close to some fitness center. We denote the set of all different SPRs over all use cases of a user u by $R_u = \bigcup_{c \in C_u} R_{u,c}$. Moreover, let $R = \bigcup_{u \in U} R_u$ be the set of all SPRs over all users.

For now, let us further assume we know values $w_{r,v} \in [0,1]$ indicating the suitability of a service point at location $v \in V$ to satisfy the needs of user $u \in U$ concerning SPR $r \in R_{u,c}$ in the use case $c \in C_u$. A value of $w_{r,v} = 1$ represents perfect suitability while a value of zero means that location v is unsuitable; values in between indicate partial suitability. For each unit of satisfied customer demand a prize $q > 0$ is earned.

A solution to the GSPDP is a subset of locations $X \subseteq V$ indicating where service points are to be set up. It is feasible if its total fixed costs do not exceed the maximum budget B, i.e.,

$$z^{\text{fix}}(X) = \sum_{v \in X} z_v^{\text{fix}} \leq B. \tag{1}$$

The objective function of the GSPDP is to maximize

$$f(X) = q \cdot \sum_{u \in U} \sum_{c \in C_u} D_{u,c} \cdot \min_{r \in R_{u,c}} \left(\max_{v \in X} w_{r,v} \right) - \sum_{v \in X} z_v^{\text{var}}. \tag{2}$$

In the first term, the obtained prize for the expected total satisfied demand is determined by considering for each user u, each use case c, and each SPR r a most suitable location $v \in V$ at which a service point is to be opened. Over all SPRs of a use case, the minimum of the obtained suitability values is taken. The second term of the objective function represents the total maintenance costs for the service stations. In [2] we have shown that the GSPDP is NP-hard.

By linearizing the objective function, the GSPDP can be modeled by the following MILP.

$$\max \quad q \cdot \sum_{u \in U} \sum_{c \in C_u} D_{u,c} \, y_{u,c} - \sum_{v \in V} z_v^{\text{var}} x_v \tag{3}$$

$$\sum_{v \in V} o_{r,v} \leq 1 \qquad\qquad\qquad \forall r \in R \tag{4}$$

$$o_{r,v} \leq x_v \qquad\qquad\qquad \forall v \in V, \, r \in R \tag{5}$$

$$y_{u,c} \leq \sum_{v \in V} w_{r,v} \cdot o_{r,v} \qquad \forall u \in U, \, c \in C_u, \, r \in R_{u,c} \tag{6}$$

$$\sum_{v \in V} z_v^{\text{fix}} x_v \leq B \tag{7}$$

$$x_v \in \{0,1\} \qquad\qquad\qquad \forall v \in V \tag{8}$$

$$0 \leq y_{u,c} \leq 1 \qquad\qquad\qquad \forall u \in U, \, c \in C_u \tag{9}$$

$$0 \leq o_{r,v} \leq 1 \qquad\qquad\qquad \forall r \in R, \, v \in V \tag{10}$$

Binary variables x_v indicate whether or not a service point is deployed at location $v \in V$. Continuous variables $o_{r,v}$ are used to indicate the actually used location $v \in V$ for each SPR $r \in R$; these variables will automatically become integer. The degree to which a use case $c \in C_u$ of a user $u \in U$ can be satisfied is expressed by continuous variables $y_{u,c}$. The objective value is calculated in (3). Inequalities (4) ensure that at most one location with the highest suitability value is selected for each SPR. Inequalities (5) and (6) ensure that use cases are only satisfied if there are suitable locations with opened service points for each SPR of the respective use case. Inequalities (6) additionally determine the degree to which a use case is satisfied. Last but not least, Inequality (7) ensures that the budget is not exceeded.

4   Cooperative Optimization Algorithm

A crucial aspect of COA's general approach is that the suitability values $w_{r,v}$ are not explicitly known a priori. A complete direct questioning would not only be extremely time consuming but users would easily be overwhelmed by the large number of possibilities, resulting in incorrect information. For example, users easily tend to only rate their preferred options as suitable and might not consider certain alternatives as also feasible although they actually might be on second thought when no other options are available.

Hence, interaction with users needs to be kept to a minimum and should be done wisely to extract as much meaningful information as possible. Therefore, COA does not ask a user to directly provide best suited station locations for the SPRs but creates meaningful *location scenarios*, i.e., subsets of locations, and asks the users to evaluate these. More specifically, a user u returns as evaluation of a location scenario S w.r.t. one of the user's SPRs $r \in R_u$ a best suited location $v_{r,S} \in S$ and the corresponding suitability value $w(r, v_{r,S}) > 0$ or the information that none of the locations of the scenario S is suitable. We assume here that the suitability of a location w.r.t. an SPR can be specified on a five valued scale.

The COA framework consists of a *Feedback Component (FC)*, an *Evaluation Component (EC)*, an *Optimization Component (OC)*, and a *Solution Management Component (SMC)*. Figure 1 illustrates the fundamental principle and communication between these components. During an initialization phase, the FC first asks each user $u \in U$ to specify her or his use cases C_u with their associated SPRs $R_{u,c}$, as well as corresponding demands $D_{u,c}$, $c \in C_u$. Then, the FC is responsible for generating individual location scenarios for each user which are presented to the user in order to obtain her/his feedback.

The obtained feedback is processed in the EC. A crucial assumption we exploit is that in a large user base some users typically have similar preferences about the locations of service points w.r.t. to some of their use cases. Hence, by identifying these similarities and learning from them, the EC maintains and continuously updates a *surrogate suitability function* $\tilde{w}_\Theta(r, v)$ approximating the real and partially unknown suitability values $w_{r,v}$ of service point locations $v \in V$ w.r.t. SPR $r \in R$ without interacting with the respective user. Based on this surrogate function, the EC also provides the surrogate objective function

$$\tilde{f}_\Theta(X) = q \cdot \sum_{u \in U} \sum_{c \in C_u} D_{u,c} \cdot \min_{r \in R_{u,c}} \left(\max_{v \in X} \tilde{w}_\Theta(r, v) \right) - \sum_{v \in X} z_v^{\mathrm{var}} \quad (11)$$

with which a candidate solution X can be approximately evaluated.

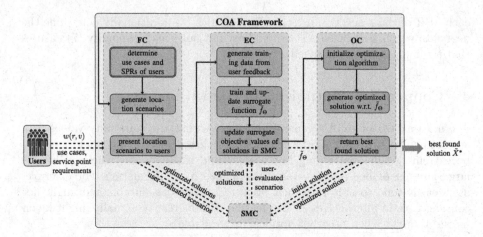

Fig. 1. Components of COA and their interaction.

A call of the OC is supposed to determine an optimal or close-to-optimal solution to the problem with respect to the EC's current surrogate objective function $\tilde{f}_\Theta$. In [2] this is achieved by solving the MILP (3)–(10) in which the suitability values are approximated by the surrogate suitability function $\tilde{w}_\Theta$.

The SMC stores and manages information on all generated solutions as well as suitability values obtained by the FC.

The whole process is repeated until some termination criterion is reached. In the end, COA returns a solution $\tilde{X}^*$ with the highest surrogate objective value of all of the so far generated solutions. For more details, in particular on how meaningful solution scenarios are derived in the FC and how a matrix factorization is utilized to determine the approximated values $\tilde{w}_\Theta(r,v)$ in the EC, we refer the interested reader to [2].

5   Large Neighborhood Search

We now propose a large neighborhood search (LNS) as a faster replacement for the original MILP-based optimization core in COA. The LNS follows the classical scheme from [15]. The key idea of LNS is to not search neighborhoods in a naive enumerative way but instead to identify via some problem-specific more effective procedure either best or promising solutions within larger neighborhoods. To this end, LNS frequently follows an iterative destroy and repair scheme: First, a given solution is partially destroyed, typically by freeing a subset of the decision variables and fixing the others to their current values. Afterwards this partial solution is repaired by finding best or at least promising values for the freed variables. If the obtained solution is better than the previous one, it is accepted, otherwise the previous solution is kept.

In our LNS a solution to a GSPDP instance is destroyed in a uniform random fashion by adding k^{dest} new locations to the solution, where k^{dest} is a parameter that is varied.

To repair a solution X, we make use of a randomized greedy approach: Let $\Delta(v, X)$ denote by how much the objective value of a solution X would decrease when removing location v from X. Note that, it is discussed later how $\Delta(v, X)$ can be efficiently calculated for all $v \in X$. In each iteration we first generate a restricted candidate list of k^{rep} locations $v \in V$ for which $\Delta(v, X)$ is lowest, i.e., the candidate list contains the locations that have the lowest impact on objective value of X. Hereby, k^{rep} is another strategy parameter. Ties are broken randomly. A location is then chosen uniformly at random from this restricted candidate list and removed from X.

To construct an initial solution in the first iteration of COA, we also make use of the repair heuristic, starting from $X = V$ and then sequentially removing locations from X for which $\Delta(v, X)$ is lowest until the solution becomes feasible, i.e. $k^{\text{rep}} = 1$ for constructing an initial solution. In subsequent iterations of COA, the LNS is warm-started with COA's current best solution $\tilde{X}^*$.

Our LNS makes use of two destroy operators with $k^{\text{dest}} = 10$ and $k^{\text{dest}} = 20$, respectively, and two repair operators with $k^{\text{rep}} = 2$ and $k^{\text{rep}} = 4$, respectively. These settings have shown to yield a robust convergence behavior across the kinds and sizes of instances in our benchmark sets. In each iteration a repair and destroy operator is chosen uniformly at random. Moreover, each LNS run terminates after 40 iterations without improvement.

A crucial aspect for developing an effective heuristic for solving the GSPDP is that computing the surrogate objective value $\tilde{f}_\Theta$ of a solution in a straight-forward way from scratch is time consuming. Hence, in order to accelerate this task we maintain for a GSPDP instance a directed graph $G = (LL \cup SL \cup CL \cup \{l_{\text{obj}}\}, A_{LL} \cup A_{SL} \cup A_{CL})$ referred to as *evaluation graph*. This graph represents the objective function calculation and stores intermediate results for a current solution, allowing for an effective incremental update in case of changes in the solution. The evaluation graph consists of four layers of nodes, which are the location layer (LL), the SPR layer (SL), the use case layer (CL), and the evaluation layer containing a single node l_{obj}. The location layer contains n nodes corresponding to the locations in V, i.e., $LL = \{l_v \mid v \in V\}$. The use case layer consists of one node for each use case C_u of each user $u \in U$, i.e., $CL = \{l_c \mid c \in C_u, u \in U\}$, and the SPR layer contains one node for each SPR in $\in R_{u,c}$, for each use case $c \in C_u$ and user $u \in U$, i.e., $SL = \{l_{u,r} \mid r \in R_{u,c}, c \in C_u, u \in U\}$.

There exists an arc in G from a node of the location layer l_v to a node of the SPR layer $l_{u,r}$ if $\tilde{w}_\Theta(v, r) > 0$, i.e., $A_{LL} = \{(l_v, l_{u,r}) \mid l_v \in LL, l_{u,r} \in SL, \tilde{w}_\Theta(v, r) > 0\}$. A node of the SPR layer is connected to a node of the use case layer if the corresponding SPR is an SPR of the corresponding use case, i.e., $A_{SL} = \{(l_{u,r}, l_c) \mid l_{u,r} \in SL, l_c \in CL, r \in R_{u,c}\}$. Finally, each node l_c of the use case layer is connected to l_{obj}, i.e., $A_{CL} = \{(l_c, l_{\text{obj}}) \mid l_c \in CL\}$.

The location layer gets as input a binary vector $(x_v)_{v \in V}$ with $x_v = 1$ if $v \in X$ and $x_v = 0$ otherwise, w.r.t. a solution X. Moreover, each node in G has an activation function $\alpha()$ that decides its output value which is propagated to its successor nodes in the next layer as their input, i.e.,

$$\alpha_{LL}(l_v, X) = \begin{cases} 1 & \text{if } v \in X \\ 0 & \text{otherwise,} \end{cases} \qquad \forall l_v \in LL, \qquad (12)$$

$$\alpha_{SL}(l_{u,r}, X) = \max_{(l_v, l_{u,r}) \in A_{LL}} (\alpha_{LL}(l_v, X) \cdot \tilde{w}_{\Theta}(v, r)) \qquad \forall l_{u,r} \in SL, \qquad (13)$$

$$\alpha_{CL}(l_c, X) = \min_{(l_{u,r}, l_c) \in A_{SL}} \alpha_{SL}(l_{u,r}, X) \qquad \forall l_c \in C_L, \qquad (14)$$

$$\alpha_{eval}(l_{obj}, X) = \sum_{(l_c, l_{obj}) \in A_{CL}} \alpha_{SL}(l_c, X) - \sum_{v \in X} z_v^{\text{var}}. \qquad (15)$$

The evaluation graph stores all output of the activation functions from the last evaluated solution and is therefore especially efficient for evaluating subsequent solutions that only differ in a single location $v \in V$ as not everything needs to be calculated from scratch but just the modified value v w.r.t. the current solution X needs to be propagated. Note that A_{LL} needs to be updated in each iteration of COA as the EC recalculates the surrogate suitability values $\tilde{w}_{\Theta}$ in each iteration with newly obtained user feedback.

Additionally, the evaluation graph also makes it possible to efficiently keep track of how much each location v contributes to the objective value of a solution. For this purpose, we introduce the following new notations. Let X be a current solution and $c \in C_u$ be a use case of a user $u \in U$ that is satisfied (to some degree) in X, i.e., for each $r \in R_{u,c}$ there exists at least one location $v \in X$ such that $\tilde{w}_{\Theta}(r, v) > 0$. Let $v^{\max}(r, X)$ refer to a location in the solution for which $\tilde{w}_{\Theta}(r, v^{\max}(r, X)) = \max_{v \in X} \tilde{w}_{\Theta}(r, v)$. For the sake of readability we further refer to $\tilde{w}_{\Theta}(r, v^{\max}(r, X))$ as $\tilde{w}_{\Theta}^{\max}(r, X)$. Additionally, let $\tilde{w}_{\Theta}^{\text{fallback}}(r, X)$ denote the second highest suitability value for an SPR r w.r.t. to the locations in X, i.e., $\tilde{w}_{\Theta}^{\text{fallback}}(r, X) = \max\{\tilde{w}_{\Theta}(r, v) \mid v \in X \setminus \{v^{\max}(r, X)\} \cup \{0\}\}$. Note that $\tilde{w}_{\Theta}^{\text{fallback}}(r, X)$ is zero if $X \setminus \{v^{\max}(r, X)\}$ is empty. Finally, let $\tilde{w}_{\Theta}^{\min}(u, c, X) = \min_{r \in R_{u,c}} \tilde{w}_{\Theta}^{\max}(r, X)$.

From the definition of the surrogate objective function, it follows that the degree to which a use case c is satisfied in a solution X is only determined by the set of locations $\{v^{\max}(r, X) \mid r \in R_{u,c}\}$. Hence, let $\Delta(u, c, v, X)$ denote by how much the degree to which a use case $c \in C_u$ of a user $u \in U$ is satisfied w.r.t. a solution X would decrease when removing v from X, i.e.,

$$\Delta(u, c, r, X) = \begin{cases} q \cdot D_{u,c} \cdot (\tilde{w}_{\Theta}^{\max}(r, X) - \tilde{w}_{\Theta}^{\text{fallback}}(r, X)) & \tilde{w}_{\Theta}^{\text{fallback}}(r, X) < \tilde{w}_{\Theta}^{\min}(u, c, X) \\ 0 & \text{otherwise} \end{cases}$$

$$\qquad (16)$$

$$\Delta(u, c, v, X) = \max\{\Delta(u, c, r, X) \mid r \in R_{u,c}, v = v^{\max}(r, X)\} \cup \{0\} \qquad (17)$$

Generally speaking, the removal of a location v from a solution X only has an impact on a use case $c \in C_u$ if it results in a change of $\tilde{w}_{\Theta}^{\min}(u, c, X)$. Additionally, note that the GSPDP also allows cases in which one service point location can be associated to multiple SPRs of the same use case. Such a case would for example correspond to situations in which a customer returns a vehicle at the same station at which the vehicle was picked up. Therefore, the removal of a location from X may affect a use case w.r.t. more than one of its SPRs. However, only the change that affects $\tilde{w}_\Theta^{\min}(u, c, X)$ the most is relevant for calculating by how much the degree to which a use case is satisfied changes.

Hence, the amount $\Delta(v, X)$ by how much the objective value of a solution would decrease when removing location v from X is calculated as

$$\Delta(v, X) = -z_v^{\text{var}} + \sum_{u \in U} \sum_{c \in C_u} \Delta(u, c, v, X). \tag{18}$$

Note that the time required for determining $w^{\max}, w^{\text{fallback}}$, and $w^{\min}$ is negligible if the domain of the rating scale by which users can specify suitability values is small. Moreover, $\Delta(v, X)$ does not need to be calculated from scratch every time a location is added or removed from the solution. Let $X \circ \{v\}$ refer to the modification of a solution, by either adding or removing a location $v \subseteq V$ to/from X. Then $\Delta(v', X \circ \{v\})$ with $v' \in X$ can be determined from $\Delta(v', X)$ as follows:

$$\Delta(v', X \circ \{v\}) = \Delta(v', X) - \sum_{u \in U} \sum_{c \in C_u} \Delta(u, c, v', X) + \Delta(u, c, v', X \circ \{v\}). \tag{19}$$

Additionally, $\Delta(v, X)$ needs to be updated only w.r.t. use cases that are actually affected by the modification of the solution, i.e., only if $\tilde{w}_\Theta^{\max}, \tilde{w}_\Theta^{\text{fallback}}$, or $\tilde{w}_\Theta^{\min}$ of a use case change. Finally, for each use case $c \in C_u$ at most $2 \cdot |R_{u,c}|$ locations need to updated in the worst case.

6   Benchmark Scenarios

Benchmark scenarios for our experiments were generated as described in detail in [2] and are available at https://www.ac.tuwien.ac.at/research/problem-instances/#spdp.

The considered test instances are of two groups. One group of instances is inspired by the location planning of car sharing systems and hence referred to as CSS. Locations are randomly generated on a grid in the Euclidean plane. The number of use cases for each user is chosen randomly, but each use case always has two SPRs. To generate suitability values for locations w.r.t. SPRs, ten *attraction points* are randomly placed on the grid, and each SPR is then associated with a geographic location sampled from a normal distribution centered around a randomly chosen attraction point. The actual suitability value is then calculated via a sigmoid function based on the distance between the SPR's geographic location and the respective service point location and afterwards perturbed by Gaussian noise. Six sets of 30 benchmark instances were generated for CSS, considering different combinations of the number of potential service point locations and the number of users.

The second group of instances is derived from real-world taxi trip data of Manhattan and referred to as MAN. The underlying street network of the instances corresponds to the street network graph of Manhattan provided by the Julia package LightOSM[1]. The Taxi trips have been extracted from the 2016 Yellow Taxi Trip Data[2]. For the generation of the instances all trips within the ten taxi zones with the highest total number of pickups and drop-offs of customers were considered, resulting in a total of approximately two million taxi trips. The set of potential service point locations has been chosen randomly from vertices of the street network that are located in the considered taxi zones. Each use case of a user is associated with two SPRs representing the origin and destination of a trip chosen uniformly at random. Suitability values for locations w.r.t. SPRs are again calculated via a sigmoid function based on the distance between the SPR's geographic location and the respective service point location. The MAN benchmark group also consists of 30 instances in total with each instance having 100 potential service point locations and 2000 users. Additionally, each instance will be evaluated with different budget levels $b\,[\%] \in \{30, 50, 70\}$ such that about b percent of the stations can be expected to be opened.

7   Computational Results

All test runs have been executed on an Intel Xeon E5-2640 v4 2.40GHz machine in single-threaded mode. Gurobi 9.1[3] was used to solve the MILP models in the OC. We compare our COA with the LNS, denoted in the following as COA[LNS], to the COA from [2] that uses the MILP (3)–(10) as optimization core and henceforth denoted as COA[MILP]. Since COA[LNS] always uses the current best solution $\tilde{X}^*$ as initial solution, we also set $\tilde{X}^*$ as starting solution in the MILP solver.

We present the results of COA by providing snapshots at different levels of performed user interactions. In [2] we have argued that at most $I_u^{\mathrm{UB}} = \sum_{r \in R_u} (|\{v \mid w(r,v) > 0\}| + 1)$ interactions per user are required to completely derive all suitability values of user $u \in U$. Let I_u be the number of user interactions of user $u \in U$ performed within COA to generate some solution. Then, $I = 100\% \cdot (\sum_{u \in U} I_u / I_u^{\mathrm{UB}})/m$, refers to the relative average number of performed user interactions relative to I_u^{UB} over all users. Results are presented in an aggregated way at various *interaction levels* ψ by selecting for each instance the COA iteration at which I is largest but does not exceed ψ.

First, we provide some general information about the performance of COA[LNS]. Table 1 shows for each instance group at different interaction levels the average number of performed destroy and repair iterations n_{iter}, the average time in seconds required for finding the best solution $t^*[s]$, and the average total time in seconds until the LNS terminated $t[s]$.

We can see that the LNS terminates within 43 to 80 iterations on average and usually terminates within three seconds for the CSS instances and within eight

---

[1] https://github.com/DeloitteDigitalAPAC/LightOSM.jl.
[2] https://data.cityofnewyork.us/Transportation/2016-Yellow-Taxi-Trip-Data/k67s-dv2t.
[3] https://www.gurobi.com/.

Table 1. Results of COA[LNS].

	CSS																
(n,m)	(100, 500)			(100, 1000)			(200, 1000)			(200, 2000)			(300, 1500)			(300, 3000)		
ψ	n_{iter}	$t^*[s]$	$t[s]$	n_{iter}	$t^*[s]$	$t[s]$	n_{iter}	$t^*[s]$	$t[s]$	n_{iter}	$t^*[s]$	$t[s]$	n_{iter}	$t^*[s]$	$t[s]$	n_{iter}	$t^*[s]$	$t[s]$
40	50	0.21	0.87	62	0.96	2.08	60	0.21	0.61	76	1.37	2.31	59	0.35	0.66	75	1.32	1.99
50	51	0.27	1.09	67	1.18	2.82	67	0.48	1.05	68	1.03	2.42	65	0.50	0.94	71	1.32	2.34
60	46	0.22	1.17	58	1.09	3.09	58	0.41	1.17	59	1.01	2.74	66	0.61	1.19	65	1.47	2.96
70	47	0.25	1.39	53	0.79	3.09	50	0.30	1.27	58	1.18	3.07	64	0.56	1.22	64	1.45	3.27
80	45	0.18	1.51	48	0.44	2.78	45	0.16	1.16	50	0.59	2.80	56	0.47	1.33	59	1.14	2.98
90	43	0.10	1.48	44	0.25	2.64	45	0.17	1.19	49	0.60	2.73	46	0.22	1.17	44	0.43	2.51

	MAN								
b	30%			50%			70%		
ψ	n_{iter}	$t^*[s]$	$t[s]$	n_{iter}	$t^*[s]$	$t[s]$	n_{iter}	$t^*[s]$	$t[s]$
40	78	2.19	3.85	74	1.58	3.35	59	0.65	1.76
50	80	3.70	6.12	75	2.76	5.25	55	1.10	3.30
60	78	4.22	8.20	72	3.72	7.21	63	2.00	5.02
70	65	3.40	8.18	64	3.10	7.74	54	1.51	5.62
80	55	2.62	7.65	58	2.93	8.12	54	1.93	6.74
90	49	1.40	7.24	48	1.27	7.35	46	0.73	6.09

seconds for the MAN instances. While the total number of iterations is relatively low, we later show in Table 2 that the solutions generated by the LNS are almost optimal w.r.t. the presented instances. The number of iterations performed tends to decrease as the number of performed user interactions increases while the total runtime increases in each iteration for the MAN instance but stays almost constant for the CSS instances. The decreasing number of iterations can be explained by the LNS being warm-started with the so far best found solution $\tilde{X}^*$. Moreover, as the number of user interactions increases, COA is able to identify more locations relevant to the SPRs of the use cases of the users, resulting in a higher number of arcs between the nodes in the service point layer and the nodes in the SPR layer of the respective evaluation graph. Therefore, the number of iterations until the LNS converges decreases while the time for performing one iteration increases.

Next, we investigate COA runs in which we apply in each iteration both, the LNS and the MILP, for solving the exact same GSPDP instances w.r.t. $\tilde{w}_\Theta$ as well as the initial solution $\tilde{X}^*$. The MILP solver is able to find optimal solution in all cases, but at the expense of typically much longer running times. Note however that only the solution generated by the LNS is further used for the next iteration in COA. Table 2 shows the average percentage gaps between the objective values of the best solutions found by the LNS and respective optimal solutions w.r.t. $\tilde{f}_\Theta$, denoted by $\text{gap}_{\tilde{f}_\Theta}[\%]$, the average total running times in seconds of the LNS $t[s]$, the average times $t^\circ_{\text{M}}[s]$ needed by the MILP solver required for reaching a solution with at most the same objective value as the solution obtained by

Table 2. Times required by the LNS, times the MILP solver needed to obtain a solution with at least the same quality as the solution of the LNS, as well as the total time required by the MILP to find a proven optimal solution. Additionally, the optimality gaps between the LNS solutions and respective optimal solutions are also shown.

	CSS											
	(100, 500)				(200, 1000)				(300, 1500)			
ψ	$t[s]$	$t_{\mathrm{M}}^{\circ}[s]$	$t_{\mathrm{M}}[s]$	$\mathrm{gap}_{\tilde{f}_\Theta}[\%]$	$t[s]$	$t_{\mathrm{M}}^{\circ}[s]$	$t_{\mathrm{M}}[s]$	$\mathrm{gap}_{\tilde{f}_\Theta}[\%]$	$t[s]$	$t_{\mathrm{M}}^{\circ}[s]$	$t_{\mathrm{M}}[s]$	$\mathrm{gap}_{\tilde{f}_\Theta}[\%]$
40	**0.87**	4.58	6.50	0.91	**0.61**	2.41	3.90	0.65	**0.66**	2.97	4.01	0.08
50	**1.09**	4.03	7.68	0.90	**1.05**	3.60	5.27	0.27	**0.94**	3.59	4.31	0.10
60	**1.17**	5.50	7.47	0.78	**1.17**	3.32	5.12	0.19	**1.19**	3.67	4.62	0.07
70	**1.39**	6.65	8.10	0.64	**1.27**	3.75	4.81	0.12	**1.22**	3.14	3.74	0.07
80	**1.51**	5.74	7.04	0.44	**1.16**	4.48	5.97	0.08	**1.33**	3.28	4.40	0.04
90	**1.48**	5.48	6.73	0.33	**1.19**	4.11	5.06	0.06	**1.17**	4.58	5.30	0.03

	CSS											
	(100, 1000)				(200, 2000)				(300, 3000)			
ψ	$t[s]$	$t_{\mathrm{M}}^{\circ}[s]$	$t_{\mathrm{M}}[s]$	$\mathrm{gap}_{\tilde{f}_\Theta}[\%]$	$t[s]$	$t_{\mathrm{M}}^{\circ}[s]$	$t_{\mathrm{M}}[s]$	$\mathrm{gap}_{\tilde{f}_\Theta}[\%]$	$t[s]$	$t_{\mathrm{M}}^{\circ}[s]$	$t_{\mathrm{M}}[s]$	$\mathrm{gap}_{\tilde{f}_\Theta}[\%]$
40	**2.08**	21.87	37.42	2.15	**2.31**	32.79	92.15	1.24	**1.99**	26.04	85.61	0.81
50	**2.82**	28.03	50.51	1.97	**2.42**	37.61	90.11	1.07	**2.34**	39.84	101.52	0.57
60	**3.09**	35.60	59.04	1.45	**2.74**	36.47	126.67	0.89	**2.96**	38.34	130.05	0.47
70	**3.09**	42.95	67.34	1.74	**3.07**	40.48	111.96	0.84	**3.27**	43.41	136.93	0.36
80	**2.78**	43.57	69.94	1.83	**2.80**	40.98	120.07	0.90	**2.98**	43.78	137.76	0.37
90	**2.64**	40.09	74.98	1.37	**2.73**	41.33	123.32	0.78	**2.51**	63.56	149.78	0.37

	MAN											
	30%				50%				70%			
ψ	$t[s]$	$t_{\mathrm{M}}^{\circ}[s]$	$t_{\mathrm{M}}[s]$	$\mathrm{gap}_{\tilde{f}_\Theta}[\%]$	$t[s]$	$t_{\mathrm{M}}^{\circ}[s]$	$t_{\mathrm{M}}[s]$	$\mathrm{gap}_{\tilde{f}_\Theta}[\%]$	$t[s]$	$t_{\mathrm{M}}^{\circ}[s]$	$t_{\mathrm{M}}[s]$	$\mathrm{gap}_{\tilde{f}_\Theta}[\%]$
40	**3.85**	67.21	326.46	2.15	**3.35**	17.24	53.54	0.87	**1.76**	6.36	10.71	0.21
50	**6.12**	80.31	328.53	1.36	**5.25**	16.76	95.43	0.59	**3.30**	10.20	15.29	0.11
60	**8.20**	131.28	368.28	1.19	**7.21**	24.36	89.15	0.43	**5.02**	14.59	21.54	0.07
70	**8.18**	140.34	375.46	1.06	**7.74**	24.86	108.59	0.35	**5.62**	13.22	21.73	0.06
80	**7.65**	160.13	414.39	1.12	**8.12**	27.70	108.01	0.34	**6.74**	18.00	24.43	0.05
90	**7.24**	154.44	411.55	1.29	**7.35**	43.43	102.70	0.27	**6.09**	13.03	17.46	0.03

the LNS, as well as the average total times $t_{\mathrm{M}}[s]$ in seconds of the MILP solver for determining a proven optimal solution. Bold values indicate best times w.r.t. $t, t_{\mathrm{M}}^{\circ}$, and t_{M}. First, we can see that the solutions generated by the LNS are on average only about 1% worse than an optimal solution for most instance groups. Next, the table shows that for CSS instances with a n/m ratio of 1/10, the MILP solver needs significantly more time for finding good solutions. Note that these instances have been designed in such a way that users behave less

Table 3. Quality of solutions generated by COA[LNS] and COA[MILP].

	CSS											
(n,m)	(100, 500)		(100, 1000)		(200, 1000)		(200, 2000)		(300, 1500)		(300, 3000)	
ψ	gap$_L$[%]	gap$_M$[%]	gap$_L$[%]	gap$_M$[%]	gap$_L$[%]	gap$_M$[%]	gap$_L$[%]	gap$_M$[%]	gap$_L$[%]	gap$_M$[%]	gap$_L$[%]	gap$_M$[%]
40	3.46	**2.98**	11.92	**9.57**	1.64	**1.21**	4.34	**2.81**	0.54	**0.51**	2.81	**2.36**
50	2.06	**1.50**	7.34	**4.72**	0.81	**0.62**	2.86	**1.90**	0.43	**0.31**	1.87	**1.27**
60	1.62	**0.63**	4.31	**2.29**	0.45	**0.36**	2.21	**1.20**	0.30	**0.20**	1.31	**0.72**
70	1.20	**0.27**	4.11	**1.61**	0.22	**0.12**	1.56	**0.47**	0.19	**0.11**	0.81	**0.28**
80	0.66	**0.18**	2.72	**0.92**	0.15	**0.05**	1.30	**0.18**	0.09	**0.04**	0.58	**0.15**
90	0.43	**0.01**	1.95	**0.08**	0.08	**0.02**	0.95	**0.05**	0.06	**0.01**	0.44	**0.03**

	MAN					
b	30%		50%		70%	
ψ	gap$_L$[%]	gap$_M$[%]	gap$_L$[%]	gap$_M$[%]	gap$_L$[%]	gap$_M$[%]
40	8.61	**3.46**	3.58	**3.46**	1.32	3.46
50	5.14	**1.88**	2.19	**1.88**	0.77	1.88
60	3.32	**1.14**	1.41	**1.14**	0.46	1.14
70	2.53	**0.63**	0.86	**0.63**	0.25	0.63
80	2.03	**0.26**	0.54	**0.26**	0.12	0.26
90	1.77	**0.10**	0.33	**0.10**	0.05	0.10

similar resulting in more complex instances. Nonetheless, the LNS significantly outperforms the MILP w.r.t. all instance groups. For all instance groups the LNS requires significantly less time on average to terminate than the MILP needs to reach a solution of the same quality as the solution obtained by the LNS. Additionally, Table 2 especially highlights how much more time the MILP requires for improving a solution at the same quality as the best found LNS solution to a provable optimal solution. Moreover, further tests have shown that most of the time the LNS is able to identify its best found solution while the MILP solver has still not yet solved the root relaxation in the same amount of time.

Finally, we want to compare independent COA[MILP] and COA[LNS] runs, and thus the impact of the in general slightly worse intermediate solutions of the LNS on the overall results of the two COA variants. For this purpose Table 3 shows for each interaction level the average optimality gaps between the best found solution during the optimization to an optimal solution w.r.t. the original objective f for COA[LNS] (gap$_L$[%]) as well as COA[MILP] (gap$_M$[%]).

The table shows that small differences in the solution quality w.r.t. $\tilde{f}_\Theta$ translate to slightly larger differences w.r.t. f. With the exception of the MAN instance group with $b[\%] = 30$, the solutions generated by COA[LNS] are usually at most 3% off from the values obtained by COA[MILP]. In most cases, the average differences are around 1% or less. Hence, in general it can be concluded that the LNS substantially outperforms the MILP in terms of computation time while still being able to generate almost optimal solutions.

8   Conclusion and Future Work

We presented a large neighborhood search (LNS) to be used as optimization core in a cooperative optimization approach (COA) for the general service point distribution problem (GSPDP) in mobility applications. While the LNS follows the traditional destroy and repair principle, a major challenge was to (a) effectively guide the repair heuristic to produce promising new solutions and to (b) efficiently calculate the surrogate objective function for modified solutions in an incremental way. Both was achieved by introducing the evaluation graph, which stores relevant intermediate results allowing efficient updates when stations are added to or removed from the current solution. In particular, the evaluation graph provides an effective way to keep track of how much impact each location in the solution has on its respective objective value. The efficient update possibility also allows to consider a larger amount of locations during the destroy procedure. The performance of the LNS within COA was tested on artificial instances as well as instances derived from real-world data and was compared to the original COA with its MILP-based optimization core. Results show that at the cost of a slight deterioration of usually not more than one percent in the quality of the solutions, the LNS can outperform the MILP w.r.t. to computation times by orders of magnitudes. In future work it seems promising to also consider other metaheuristic approaches, such as an evolutionary algorithm that uses the evaluation graph for efficiently recombining solutions. Moreover, the GSPDP it is still a rather abstract problem formulation, and it would be important to extend it as well as the solving approach to cover further relevant practical aspects such as capacities of stations and time dependencies of users.

References

1. Jatschka, T., Rodemann, T., Raidl, G.R.: A cooperative optimization approach for distributing service points in mobility applications. In: Liefooghe, A., Paquete, L. (eds.) EvoCOP 2019. LNCS, vol. 11452, pp. 1–16. Springer, Cham (2019). https://doi.org/10.1007/978-3-030-16711-0_1

2. Jatschka, T., Raidl, G., Rodemann, T.: A general cooperative optimization approach for distributing service points in mobility applications. Technical report AC-TR-21-006, TU Wien, Vienna, Austria (2021, submitted)

3. Jatschka, T., Rodemann, T., Raidl, G.R.: VNS and PBIG as optimization cores in a cooperative optimization approach for distributing service points. In: Moreno-Díaz, R., Pichler, F., Quesada-Arencibia, A. (eds.) EUROCAST 2019. LNCS, vol. 12013, pp. 255–262. Springer, Cham (2020). https://doi.org/10.1007/978-3-030-45093-9_31

4. Meignan, D., Knust, S., Frayret, J.M., Pesant, G., Gaud, N.: A review and taxonomy of interactive optimization methods in operations research. ACM Trans. Interact. Intell. Syst. **5**, 17:1–17:43 (2015)

5. Sun, X., Gong, D., Jin, Y., Chen, S.: A new surrogate-assisted interactive genetic algorithm with weighted semisupervised learning. IEEE Trans. Cybern. **43**, 685–698 (2013)

6. Sun, X.Y., Gong, D., Li, S.: Classification and regression-based surrogate model-assisted interactive genetic algorithm with individual's fuzzy fitness. In: Proceedings of the 11th Annual Conference on Genetic and Evolutionary Computation, pp. 907–914. ACM (2009)

7. Llorà, X., Sastry, K., Goldberg, D.E., Gupta, A., Lakshmi, L.: Combating user fatigue in iGAs: partial ordering, support vector machines, and synthetic fitness. In: Proceedings of the 7th Annual Conference on Genetic and Evolutionary Computation, pp. 1363–1370. ACM (2005)

8. Kim, H.S., Cho, S.B.: Application of interactive genetic algorithm to fashion design. Eng. Appl. Artif. Intell. **13**, 635–644 (2000)

9. Dou, R., Zong, C., Nan, G.: Multi-stage interactive genetic algorithm for collaborative product customization. Knowl.-Based Syst. **92**, 43–54 (2016)

10. Jatschka, T., Rodemann, T., Raidl, G.R.: Exploiting similar behavior of users in a cooperative optimization approach for distributing service points in mobility applications. In: Nicosia, G., Pardalos, P., Umeton, R., Giuffrida, G., Sciacca, V. (eds.) LOD 2019. LNCS, vol. 11943, pp. 738–750. Springer, Cham (2019). https://doi.org/10.1007/978-3-030-37599-7_61

11. Bell, R.M., Koren, Y., Volinsky, C.: Matrix factorization techniques for recommender systems. Computer **42**, 30–37 (2009)

12. Ekstrand, M.D., Riedl, J.T., Konstan, J.A.: Collaborative filtering recommender systems. Found. Trends Hum.-Comput. Interact. **4**, 81–173 (2011)

13. Frade, I., Ribeiro, A., Gonçalves, G., Antunes, A.: Optimal location of charging stations for electric vehicles in a neighborhood in Lisbon, Portugal. Transp. Res. Rec.: J. Transp. Res. Board **2252**, 91–98 (2011)

14. Kloimüllner, C., Raidl, G.R.: Hierarchical clustering and multilevel refinement for the bike-sharing station planning problem. In: Battiti, R., Kvasov, D.E., Sergeyev, Y.D. (eds.) LION 2017. LNCS, vol. 10556, pp. 150–165. Springer, Cham (2017). https://doi.org/10.1007/978-3-319-69404-7_11

15. Gendreau, M., Potvin, J.Y., et al.: Handbook of Metaheuristics, vol. 3. Springer, Cham (2019). https://doi.org/10.1007/978-3-319-91086-4

A Physarum-Inspired Approach
for Influence Maximization

Álvaro O. López-García⬭, Gustavo Rodríguez-Gómez(✉)⬭,
and Aurelio López-López⬭

Instituto Nacional de Astrofísica, Óptica y Electrónica, Luis Enrique Erro No. 1,
Sta. María Tonantzintla, 72840 Puebla, Mexico
{alvarolopez,grodrig,allopez}@inaoep.mx

Abstract. The influence maximization problem (IM) is an open problem in graph theory, and also is identified as a NP-hard, so there has
been a lot of developments in order to solve or approximate a solution.
In this paper, we present an approach for pointing to a solution of the IM
problem, by leveraging k-shell decomposition analysis, and combining it
with a *Physarum*-inspired model. Additionally, this procedure was tested
on five data-sets both synthetic and real, showing encouraging results.

Keywords: Physarym polycephalum · Influence maximization · Social
networks

1 Introduction

Social media is a global phenomenon that changed our lives forever, altering the
way we perceive the world and also has an important influence in our opinions,
ideas, and decisions. They are an important and fundamental way of expressing
ourselves to the world. For 2020, it is estimated that 3.8 billion of users worldwide
are using actively social media, that is the 84% of overall web population [1].

Studying social media is an important field of interest to researchers of various
areas such as sociology, psychology, mathematics and computer science. The
study of social media is important because this is an information goldmine for
advertisers and reaches a very wide audience, providing additionally a lot of
personal user data.

One successful strategy adopted for dissemination of information and products are the so-called *influencers*. They are Internet personalities, quite popular
with a lot of followers, and consequently called "opinion leaders". The concept
of influencer is created around the idea of influential marketing, which identifies
people with a lot of influence over potential buyers, and all the marketing is
constructed around these influencers [2].

From a computer science viewpoint, the identification of influencers is an
instance of a bigger problem, called influence maximization on graphs (IM). This
problem is an open problem in graph theory [7], and is identified as a NP-hard
problem (a problem whose solution can not be computed with a polynomial-time

© Springer Nature Switzerland AG 2022
B. Dorronsoro et al. (Eds.): META 2021, CCIS 1541, pp. 18–32, 2022.
https://doi.org/10.1007/978-3-030-94216-8_2

algorithm), so there have been several proposals in this area, with new algorithms and heuristics for approximating a solution [3]. Note that those proposals have limitations, either with the size of the instances that they can solve or with the graph topologies.

One of the newest and interesting approaches is that based on a biological being and its behavior in the natural habitat. This organism is *Physarum Polycephalum*, and scientists are interested on it because of its intelligent behavior, lack of a central brain, and also given that can solve a maze to find food and nutrients [4]. Also, it can optimize energy consumption when transfering nutrients in its body, so it can solve problems like the optimal traffic network problem [5].

In this paper, we present an approach for solving the IM problem, by leveraging k-shell decomposition analysis, and combining it with a *Physarum Polycephalum* model. The k−shell decomposition is an useful technique because it performs an initial selection of nodes with a high degree (and potentially high propagation ability), and then the *Physarum Polycephalum* model is employed for evaluating nodes and determines which nodes are the most influential.

The contributions of this paper are: a new procedure for obtaining the influential nodes in a social network graph by using k−shell decomposition in conjunction with a Physarum model. Additionally, this procedure is tested on both synthetic and real data-sets. This is of importance as a part of an approach for solving the IM problem using a bio-inspired algorithm.

This paper is organized as follows. Section 2 shows related work for both the IM problem and developments with *Physarum Polycephalum*. In Sect. 3, we show the theoretical basis of the IM problem, k−shell decomposition, the Physarum model, and the degree index. The proposed method is shown and expanded on Sect. 4. Section 5 details the data-sets, experiments and results of our approach. Finally, Sect. 6 includes our conclusions and future work.

2   Related Work

When first computer networks were created, their high potential for communicating with people was perceived, since distance was no longer a limitation for communicating with friends and relatives overseas. In particular, the massification of the Internet in 1995 [6] catalyzed the development of new online platforms for communicating with friends, one of them was the social networks.

A social network service is an Internet platform, employed for creating networks or relationships among people with similar interests. In addition, such platforms allow to create friend lists and meet new people. Examples of social networks are Facebook, Twitter, QQ, or TikTok. A social network can be modeled as a graph, since this is based on establishing relations among people, and they, at the same time, can also have their own friends. So, we can build a huge graph of people connected by their relationships. Each node is a person, and each vertex represents a given relation (e.g. friendship, or con-generic) between two. Open problems in social networks [7] are: community detection, recommendation systems, trust prediction, opinion mining, and influence maximization.

Influence maximization (IM) was formulated by Kempe et al. [3], as a combinatory optimization problem (i.e., we need to search a near object from a finite object set), also proposing a greedy algorithm for solving it. This algorithm is initialized with an empty seed set, and then searches for nodes that maximizes influence. The downside of this algorithm is that on big graphs, computing time grows fast, because of the number of calculations needed. Kempe et al. identified two diffusion models: Independent Cascade (IC) and Lineal Threshold (LT) [8].

Leskovec et al. [9] proposed a new algorithm called Cost Effective Lazy Forward (CELF), and it promised faster speed (marginally) on solving the problem, compared to the traditional greedy algorithm. This algorithm is based on the modularity function of the diffusion model, so it can select nodes faster and more precisely. Also, the algorithm can prevent unnecessary calculations for diffusion, so is faster than the traditional greedy algorithm [10]. CELF also has its limitations, so Goyal et al. [11] proposed an improvement called CELF++, showing an increase of 35–55% of performance.

On other line, scientists have been trying to solve IM from an heuristic approach with algorithms such as Local Directed Acyclic Graph (LDAG), proposed by W. Chen et al. [12] This algorithm only works with LT diffusion model, but the authors argue that an improvement in time is achieved, decreasing it from hours to seconds (or from days to minutes).

Other important heuristic is PageRank, developed by Larry Page for the Google search engine [13], adapted for the IM problem by Li Q et al. [14], calling it Group-PageRank. This algorithm only works on graphs with IC diffusion model. The original idea of PageRank is that each node of the graph is given a score, according to a probability of being activated by a user on the web [10].

The *Physarum Polycephalum* approach is a recent development in bio-inspired computation. Adamatzky et al. [16] proposed one of the earliest problems solved by this model, which was facing the optimization of traffic network. But also *Physarum Polycephalum* model has been explored for solving other problems such as solving mazes [17], the Steiner tree problem in networks [18], the graph coloring problem [19], among other graph-related tasks [20].

As part of our bio-inspired optimization Physarum-based approach, there are some related works, such as that proposed by Gao C. et al. [21], who developed a new method for obtaining the centrality degree of a node, which is important for establishing how likely a node is going to be influential in a social network. This new centrality degree is called Physarum centrality, based on properties of *Physarum Polycephalum*, and also is supported by $k-$shell decomposition for identifying nodes. This algorithm works on weighted and unweighted networks.

3   Theoretical Framework

A social network is a structure formed with entities and a group of 2−way links among them (for example, friendship or family relationships). A social network can be modeled as a directed or undirected graph G, weighted or unweighted.

3.1   Influence Maximization

An open and important problem in graph theory is influence maximization (IM) [7]. This problem has been studied and showed to be NP-Hard [10], indicating that is a problem that does not have a verification algorithm in polynomial time.

A social network is studied and represented as a graph $G = (V, E)$, where V is the set of nodes in G (e.g. users), E is the set of edges in G (the relationships between the users). The aim is to find the set of users with the maximum influence in G [10].

IM is an optimization problem, which consists in maximizing the spread of information or influence in a social network graph. Formally speaking [10]:

Definition 1. *Given a social graph $G = (V, E)$ and a user set $S \subseteq V$, a diffusion model M captures the stochastic process for S spreading information on G. Influence spreading of S, denoted as $\sigma_{G,M}(S)$, is the number of expected users influenced by S.*

Definition 2. *Given a social graph $G = (V, E)$, a diffusion model M and a positive integer k, influence maximization selects a set S^* of k users of V as a seed set for maximizing influence diffusion, such as $S^* = \arg\max_{S \subseteq V \wedge |S| \leq k} \sigma_{G,M}(S)$.*

For the IM problem there are various diffusion models, but the most common are Independent Cascade (IC) and Lineal Threshold (LT). The aim of these models is to associate each user in G a status (i.e. active or inactive) and the conditions of activating or infecting them. Independent Cascade (IC) states that there is a probability of infection for each edge, namely P_{ij}, where P is the probability of i infecting j. Once j is infected, it can infect neighbours on the next step, according to the probability assigned to next edge. Linear Threshold (LT) is different, because each node is infected if neighbour nodes are infected by reaching a threshold, according to their weights [22].

3.2   K-shell Decomposition

K-shell decomposition is a technique for decomposing and studying the structure of large graphs. This method is also noted for showing the importance of certain nodes in regards of their hierarchies.

The following concepts are basic for this decomposition, as given by [23].

Definition 3. *The k-Core of a graph G is the maximal subgraph of G having minimum degree at least k.*

Definition 4. *The k-Shell of a graph G is the set of all nodes belonging to the k-Core of G but not to the (k+1)-Core.*

The k-shell index of a node is denoted as K_s.

Kitsak et al. [24] proposed the use of k-shell decomposition as a means for identifying node spreaders in a graph network. So, the higher the k-shell index of a node, the more it can spread information in the network. The k-shell decomposition works better on static networks, where topologies do not change over time [23].

3.3 Physarum Model

Our model for *Physarum Polycephalum* is based on the works of Z. Cai et al. [5] and Gao C. [21]. This model simulates the foraging behavior of *Physarum Polycephalum*. Its body is a single cell made up of interconnected tubes forming networks, that can stretch from centimeters to meters, and can store and recover information when searching for food. [15]. On laboratory experimental setups, the model consists of a Petri dish, a map (usually made of agar), external food sources (usually oat flakes) and a live Physarum [16].

Physarum Polycephalum consists of the following components [5]:

1. **Plasmodium and Myxamoebas**: The plasmodium is the moving part of the organism, and the tentacle-shaped myxamoebas are the deformed part of the plasmodium. They are used for foraging and consuming food and nutrients, and expand and contract accordingly.
2. **Nucleus**: The nucleus is the central and critical part of the organism, which moves and feeds around it.
3. **Nutrients**: They are the source of energy, and come from external food sources.

These all parts work collaboratively in order to solve problems, and the solving process includes two stages: food searching, and feeding [5]. The first stage is when the myxamoebas start growing around the nucleus in order to find external food sources. The second stage is when the myxamoebas contract in order to transport all the found nutrients in its body. In terms of optimization, these two stages correspond to exploration and exploitation of the search space.

Multiple myxamoebas m grow around the nucleus, by expanding in multiple directions [5]. This growth is constrained by the topology of its environment, and in our case the topology of the social network graph, i.e. its adjacency matrix, which is a time-varying structure, denoted as follows:

$$\mu = [\mu_{ij}(t)]_{n \times n} \tag{1}$$

where $\mu_{ij} = 1$ when there is a direct edge from node i to j, or $\mu_{ij} = 0$ otherwise.

Another important part of our model is the nutrient concentration matrix on the edges, which is also a time-varying structure, and defined as [5]:

$$\tau(t) = [\tau_{ij}(t)]_{n \times n} \tag{2}$$

There are two operations related to nutrient consumption, namely the enhancing operation ($\Delta > 0$) and the decreasing operation ($\sigma > 0$). The first one is used to simulate the nutrient transportation through the Physarum body, and the latter is intended to simulate the nutrient consumption by other life activities [5]. The nutrient concentration on each edge is updated at time t and m number of myxamoebas, with the following formula:

$$\tau_{ij}(t) = \tau_{ij}(t-1) + \sum_{k=1}^{m} \mu_{ij}^{k}(t)\Delta_{ij}^{k}(t) - \sum_{k=1}^{m} \sigma_{ij}^{k} \tag{3}$$

3.4   Node Selection by Propagation Capability

One detail to observe is that node selection is correlated with identifying the importance of each node or edges on the social network graph. There are various node indices for doing this selection, namely the degree of a node, the importance degree index, the closeness index, the betweenness degree index, and the redundancy rate index, among others [22].

In our research, we employ the betweenness index, since this considers the myxamoebas passing through node i and reaching outer nodes, and that is critical for information spreading. Betweenness measures the extent to which a node lies in the path between others, so it can measure the influence a node has over the spread of information through the network [28]. The betweenness index takes into account the number of myxamoebas passing through node i, so the more of them go by, the more important node i is.

Let λ_{jk} be the number of myxamoebas from node j to k passing through a node i. The betweenness index Bt_i of a node i can be calculated as [26]:

$$Bt_i = \sum_{j=1}^{n}\sum_{k=1}^{n} \lambda_{jk} \tag{4}$$

In a social network with n nodes, at most $n-1$ neighbor nodes can connect to a node i, so the betweenness degree index of a node i is calculated by:

$$\overline{Bt_i} = \frac{Bt_i}{\sum_{j=1}^{n} Bt_j} \tag{5}$$

In consequence, the total betweenness index of all the myxamoebas can be calculated by:

$$Bt_i^m = \sum_{k=1}^{m} \overline{Bt_i} \tag{6}$$

4   Proposed Method

Our proposed method follows most of the steps proposed by Gao C. et al. [21], in which they apply a Physarum model based on a Poisson equation, and after that, they use $k-$shell decomposition for calculating the called Physarum Centrality. However, we employ the method used by Z. Cai et al. [5], in which the most important part of the model are the growth of the myxamoebas and the nutrient consumption by the organism.

The first step is to initialize the procedure, having the social network graph $G = (N, E)$ as the only input parameter. So the adjacency matrix μ is defined in this first step (see expression (1)). Then, thereafter there is a $n \times n$ matrix, with n being the total number of nodes in the social network graph G.

The nutrient matrix concentration τ at starting time is a zero matrix of size $n \times n$ (with n the total number of nodes in the social network graph G), since at the start time there is no nutrient flowing through the body of the organism.

The increment (Δ) and decrement (σ) parameters are also initialized, with the constraints of $\Delta > 0$, $\sigma > 0$, and following the rule of thumb $\Delta > \sigma$, given that the nutrients consumed should be higher than the nutrients consumed for growing and foraging.

Once all the parameters are initialized, we proceed to compute the $k-$shell decomposition on the social graph G, for obtaining the K_s value for each node. This process basically consists on the following [25]:

```
def kShell(G):
    h = G.copy()
    it = 1
    tmp = []
    buckets = []
    while (1):
        flag = kShell_check(h, it)
        if (flag == 0):
            it += 1
            buckets.append(tmp)
            tmp = []
        if (flag == 1):
            node_set = kShell_find_nodes(h, it)
            for each in node_set:
                h.remove_node(each)
                tmp.append(each)
        if (h.number_of_nodes() == 0):
            buckets.append(tmp)
            break
    return buckets
```

Once each node has its K_s value, the nodes with higher index value are selected to function as nucleus and grow myxamoebas, so the next step is grow the myxamoebas on the whole social network graph G. Each nucleus can grow m number of myxamoebas on it, depending on the topology of the social network graph G. This m number of myxamoebas for each nucleus can be determined as follows:

$$m = \begin{cases} K_s, & \text{if } K_s < 5 \\ \left\lceil \frac{K_s}{2} \right\rceil, & \text{otherwise} \end{cases} \qquad (7)$$

where K_s is the $k-$shell decomposition index value of the selected node acting as the nucleus. The idea is to explore as many connecting paths as possible when the nodes are not highly connected but administer resources when the node is heavily connected.

For growing the myxamoebas, a recursive function is applied, with the following parameters: the adjacency list μ, the node acting as nucleus, the number of m myxamoebas, and an array used for keeping track of the visited nodes. This function is defined as follows:

```
def growMyxo(adjList, node, totalMyx, visited=[]):
    neighbours = getNeighbours(node, adjList)
    if (totalMyx>=len(neighbours)):
        selected_nodes = neighbours
    else:
        selected_nodes = random.sample(neighbours, totalMyx)
    for i in range(0,len(selected_nodes)):
        if (not selected_nodes[i] in visited):
            visited.append(selected_nodes[i])
            growMyxo(adjList, selected_nodes[i], totalMyx,
                visited)
    return visited
```

This function returns all the visited nodes by the myxamoeba while searching for food, and the nutrient matrix is updated according to the expression (3).

The feeding stage of the organism consists of a loop iterating over the m myxamoebas while updating the nutrient concentration matrix with (3). The more myxamoebas pass through a node, the higher concentration of nutrients this will have, and that will be reflected on the nutrient matrix τ. After the feeding stage, the betweenness degree index of each node of the myxamoebas grown on the selected nucleus will be calculated using expression (5).

For output of the procedure, the total betweenness degree index for all the nodes on the social network graph G is calculated by formula (6).

5   Experiments and Results

Five experiments were done in order to validate the proposed approach, using two synthetic examples and three real data-sets from social network graphs.

For running the experiments, the number of myxamoebas assigned to each social graph was defined by (7). For each data-set, the particular defined value is detailed later on. The total number of iterations for the feeding stage was set to 10. Since each myxamoeba grows differently in each run, the method was run ten times, and then average and standard deviation were calculated.

5.1   Synthetic Data-Sets

The first synthetic graph is based on the graph used in [21] and this has 15 nodes and 21 edges, with a density value of 0.2. The reported maximum degree of a node is 6, shown in Fig. 2. For this graph, the m value was set to 3.

To illustrate the process of growing myxamoebas to explore the first graph, Fig. 1 includes two examples of sub-graphs obtained when taking two different nodes as nucleus.

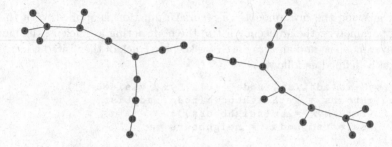

Fig. 1. Examples of growing myxamoebas on graph of Fig. 2. The graph on the left had node 11 as nucleus, and the graph on the right took node 5 as nucleus.

The results of running our method on the first synthetic graph are shown on Table 1. For comparison, we report also the Physarum centrality index (Ck_p) as described in [21]. The Top-4 nodes selected by the Physarum centrality are the same as those selected by our method, showing that nodes 7, 11, 12 and 5 have the potential of spreading the information the most in the social network graph. Also the table shows that the nodes with the least propagation ability are again the same (nodes 0, 1, 3 and 2).

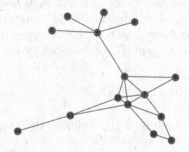

Fig. 2. Network with 15 nodes and 21 edges. After k-shell decomposition ($K_s = 3$), nodes 5, 7, 11 and 12 are selected

The second synthetic graph is based on the graph employed in [8], having 8 nodes and 20 edges, with a density value of 0.444. The reported maximum degree of a node is 8, as shown in Fig. 3. For this second synthetic data-set, the value of m was set to 2.

Table 1. Selected nodes for graph shown in Fig. 2 ordered by betweenness degree index (B.d.i.)

Node	B.d.i. (avg.)	Std deviation	Ck$_p$
5	0.350	0.018	0.265
7	0.350	0.018	0.113
11	0.350	0.018	0.182
12	0.350	0.018	0.093
8	0.347	0.018	0.052
9	0.347	0.018	0.031
10	0.340	0.019	0.038
13	0.320	0.017	0.064
14	0.310	0.023	0.011
6	0.241	0.029	0.024
4	0.196	0.029	0.084
0	0.177	0.026	0.011
1	0.133	0.026	0.011
3	0.114	0.024	0.011
2	0.077	0.035	0.011

The results of running our method on the second graph are shown in Table 2. As we can notice, nodes 0 and 1 are reported as those having the best capability to spread information, and they are also the nodes selected by the CELF algorithm in [8]. The topology of this graph shows that nodes 0 and 1 are the best spreaders, since this is pretty clear because they are in the center and have the most outer connected nodes.

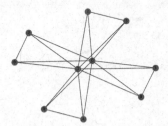

Fig. 3. Network with 8 nodes and 20 edges. During the k-shell decomposition, all nodes (0, 1, 2, 3, 4, 5, 6, 7 and 8) are selected with $K_s = 2$

Table 2. Selected nodes for graph shown in Fig. 3 sorted by degree index (B.d.i.)

Node	B.d.i. (avg.)	Std deviation
0	1.086	0.018
1	1.086	0.018
4	1.024	0.026
8	1.024	0.026
2	0.986	0.038
6	0.971	0.024
3	0.933	0.025
7	0.986	0.038
5	0.971	0.024
9	0.933	0.025

5.2 Real Data-Sets

The first real data-set is the Zachary's karate club network of 1977, containing the social ties between the members of a university karate club [27]. This graph consists of 34 nodes, 78 edges and a density value of 0.139037. The reported maximum degree of a node is 17. This graph is shown in Fig. 4. Using k-shell decomposition nodes 1, 2, 3, 4, 8, 9, 14 31, 33 and 34 are selected with $K_s = 3$. In consequence, the value for m was set to 3.

The Top-10 results for the karate club network are shown in Table 3, and these are similar to those reported in [21]. They show that nodes 1, 34, 3, 33 and 14 (in bold in the table) have the greatest Physarum centrality index, while our method shows the same nodes (except by node 14) in addition of nodes 2 and 4, are those with highest betweenness degree index.

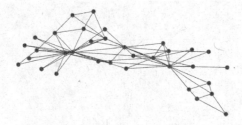

Fig. 4. Zachary's karate club social network graph.

The second real data-set is the bottlenose dolphins social network, containing a list of links, where each link is a frequent association between dolphins [27]. This graph consists of 62 nodes, 159 edges and a density value of 0.0840825. The reported maximum degree of a node is 12. This graph is shown in Fig. 5. After

Table 3. Top-10 nodes for the karate club graph, sorted by betweenness degree index (B.d.i.)

Node	B.d.i. (avg.)	Std deviation
34	0.403	0.012
33	0.403	0.015
2	0.403	0.012
3	0.403	0.012
4	0.403	0.015
1	0.403	0.006
28	0.389	0.029
24	0.389	0.030
26	0.374	0.017
32	0.374	0.034

k-shell decomposition, thirty six nodes were selected with $K_s = 3$ (i.e. nodes 1, 2, 7, 8, 9, 10, 11, 14, 15, 16, 17, 18, 19, 20, 21, 22, 25, 29, 30, 31, 34, 37, 38, 39, 41, 42, 43, 44, 46, 48, 51, 52, 53, 55, 58 and 60), so also for this graph the value for m was set to 3.

This data-set, the dolphin social network graph, showed an interesting behavior. The results of running our method on this graph are summarized in Table 4. After growing the myxamoebas on all the thirty six nodes selected after the k-shell decomposition, acting as nucleus and performing the feeding stage, the list of possible influential nodes narrowed down to nodes 15, 21 and 46. An implementation of the CELF algorithm applied on this data-set, selected nodes 15 and 46.

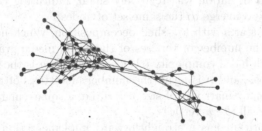

Fig. 5. Bottlenose dolphins social network graph.

The third real dataset is the public figures network, gathered from Facebook [27]. This graph consists of 11.6K nodes, 67K edges and a density value of 0.00100253. The reported maximum degree of a node is 326. Using k-shell decomposition, 170 nodes were selected with $K_s = 41$. In consequence, the value for m was set to 22.

Table 4. Top-10 nodes for the dolphin graph, sorted by betweenness degree index (B.d.i.)

Node	B.d.i. (avg.)	Std deviation
15	0,719	0,001
21	0,719	0,001
46	0,719	0,001
38	0,718	0,004
30	0,714	0,003
34	0,713	0,007
43	0,712	0,006
48	0,712	0,008
41	0,711	0,008
39	0,711	0,010

The implementation of the CELF algorithm applied to this data-set, selected 191 nodes. Our method selected 2963 nodes as top ranked, with a recall of 122 (64%) of those selected by CELF. The range of standard deviation is between 0.006 and 0.000004. Among the 10 runs of our proposed algorithm, the ninth showed the best behavior and the second showed the worst. It is important to note that because of the non-deterministic nature of the approach, this might vary over time.

5.3   Discussion

The experiments on synthetic data-sets allow to verify that the approach was working adequately. The further validation on real data-sets led to similar results as previously reported or as identified by a previous algorithm. In all the data-sets, the standard deviation was relatively small, indicating that most of the executions tend to converge to the same set of nodes.

The approach starts with a k-shell decomposition, which has a complexity of $O(n)$, with n the number of vertices of the social media graph. Overall the proposed approach has a complexity of $O(\mu n^3)$, where μ is the total number of grown myxamoebas, and n is again the number of vertices of the social media graph. The adjacency matrix is of size $n \times n$, i.e. a square matrix representing the edges between all the n vertices.

Also, our algorithm has a straightforward implementation for the experiments, so we employed big matrices for doing all the computation, and this constrains the size of the data-sets. One improvement can be handling graphs as linked lists, which would demand less memory and work with larger data-sets.

6   Conclusion and Future Work

In this paper, a method for obtaining the influential nodes in a social network graph is proposed, by using k-shell decomposition and by simulating the behavior of a *Physarum Polycephalum*. The growth of its myxamoebas and its nutrient consumption are illustrated and used in our method, in order to filter which nodes are the most influential. After simulating the food searching and feeding stage, the method will help determine which nodes are the most influential in the social network graph. For this, the betweenness degree index and the myxamoebas that pass through a particular node are considered. The experiments showed that the approach was working well and reached similar results as those reported in earlier works.

The proposed method has to be further tested on larger graphs, but since IM is a NP-hard problem, we might need to recur to improved computing infrastructure, in order to operate on such graphs. Also a further complexity analysis has to be done, in order to have a better idea of the efficiency and applicability of the proposed method.

Acknowledgements. First author thanks the support of Conacyt, México through scholarship number 1008092. The other authors had partial support of SNI.

References

1. We are Social, Hootsuite. Digital in 2020, January 2020. https://wearesocial.com/digital-2020
2. Agencia de Publicidad Endor. Qué es un Influencer y cuál es su función?, 5 March 2020. https://www.grupoendor.com/influencer-funcion/. Accessed 1 July 2020
3. Kempe, D., Kleinberg, J., Tardos, E.: Maximizing the Spread of Influence through a Social Network. ACM (2003). https://www.cs.cornell.edu/home/kleinber/kdd03-inf.pdf
4. Tang, C.-B., Zhang, Y., Wang, L., Zhang, Z.: What can AI learn from bionic algorithms? Phys. Life Rev. **29**, 41–43 (2019). https://doi.org/10.1016/j.plrev.2019.01.006
5. Cai, Z., Xiong, Z., Wan, K., Xu, Y., Xu, F.: A node selecting approach for traffic network based on artificial slime mold. IEEE Access **8**, 8436–8448 (2020). https://doi.org/10.1109/access.2020.2964002
6. Schafer, V., Thierry, B.G.: The 90s as a turning decade for Internet and the Web. Internet Histories **2**(3–4), 225–229 (2018). https://doi.org/10.1080/24701475.2018.1521060
7. Problems in Graph Theory and Combinatorics. (n.d.). https://faculty.math.illinois.edu/. Accessed 1 July 2020
8. Kingi, H.: Influence Maximization in Python-Greedy vs CELF, 7 September 2018. https://hautahi.com/im_greedycelf. Accessed 29 June 2020
9. Leskovec, J., Krause, A., Guestrin, C., Faloutsos, C., Vanbriesen, J., Glance, N.: Cost-effective outbreak detection in networks. In: Proceedings of the ACM SIGKDD International Conference on Knowledge Discovery and Data Mining, pp. 420–429 (2007). https://doi.org/10.1145/1281192.1281239
10. Li, Y., Fan, J., Wang, Y., Tan, K.L.: Influence maximization on social graphs: a survey. IEEE Trans. Knowl. Data Eng. **30**(10), 1852–1872 (2018). https://doi.org/10.1109/tkde.2018.2807843
11. Goyal, A., Lu, W., Lakshmanan, L.V.S.: CELF++. In: Proceedings of the 20th International Conference Companion on World Wide Web WWW 2011 (2011)

12. Chen, W., Yuan, Y., Zhang, L.: Scalable influence maximization in social networks under the linear threshold model. In: 2010 IEEE International Conference on Data Mining, Sydney, NSW, pp. 88–97 (2010). https://doi.org/10.1109/ICDM.2010.118

13. Page, L., Brin, S., Motwani, R., Winograd, T.: The PageRank citation ranking: bringing order to the web. Stanford University Info Lab, pp. 1–17 (1998). http://ilpubs.stanford.edu:8090/422/

14. Liu, Q., Xiang, B., Chen, E., Xiong, H., Tang, F., Yu, J.X.: Influence maximization over large scale social networks. In: Proceedings of the 23rd ACM International Conference on Information and Knowledge Management, CIKM 2014, pp. 171–180 (2014). https://doi.org/10.1145/2661829.2662009

15. Staff, S.X.: Researchers find a single-celled slime mold with no nervous system that remembers food locations. Phys. Org., 23 February 2021. https://phys.org/news/2021-02-single-celled-slime-mold-nervous-food.html

16. Adamatzky, A., Jones, J.: Road planning with slime mold: if physarum built motorways it would route M6/M74 through newcastle. Int. J. Bifurcat. Chaos **20**(10), 3065–3084 (2010). https://doi.org/10.1142/s0218127410027568

17. Tero, A., Kobayashi, R., Nakagaki, T.: Physarum solver: a biologically inspired method of road-network navigation. Phys. A: Stat. Mech. Appl. **363**(1), 115–119 (2006). https://doi.org/10.1016/j.physa.2006.01.053

18. Liu, L., Song, Y., Zhang, H., Ma, H., Vasilakos, A.V.: Physarum optimization: a biology-inspired algorithm for the steiner tree problem in networks. IEEE Trans. Comput. **64**(3), 818–831 (2015). https://doi.org/10.1109/tc.2013.229

19. Lv, L., Gao, C., Chen, J., Luo, L., Zhang, Z.: Physarum-based ant colony optimization for graph coloring problem. In: Tan, Y., Shi, Y., Niu, B. (eds.) ICSI 2019. LNCS, vol. 11655, pp. 210–219. Springer, Cham (2019). https://doi.org/10.1007/978-3-030-26369-0_20

20. Zhang, X., Mahadevan, S., Deng, Y.: Physarum-inspired applications in graph-optimization problems. Parallel Process. Lett. **25**(01), 1540005 (2015). https://doi.org/10.1142/s0129626415400058

21. Gao, C., Lan, X., Zhang, X., Deng, Y.: A bio-inspired methodology of identifying influential nodes in complex networks. PLoS ONE **8**(6), e66732 (2013). https://doi.org/10.1371/journal.pone.0066732

22. Shakarian, P., Bhatnagar, A., Aleali, A., Shaabani, E., Guo, R.: The independent cascade and linear threshold models. In: Diffusion in Social Networks. SCS, pp. 35–48. Springer, Cham (2015). https://doi.org/10.1007/978-3-319-23105-1_4

23. Bickle, A.: The k-Cores of a Graph. Dissertations. 505 (2010). https://scholarworks.wmich.edu/dissertations/505

24. Kitsak, M., et al.: Identification of influential spreaders in complex networks. Nat. Phys. **6**(11), 888–893 (2010). https://doi.org/10.1038/nphys1746

25. GeeksforGeeks. K-shell decomposition on Social Networks, 1 October 2020. https://www.geeksforgeeks.org/k-shell-decomposition-on-social-networks/

26. Bauer, B., Jordán, F., Podani, J.: Node centrality indices in food webs: Rank orders versus distributions. Ecol. Complex. **7**(4), 471–477 (2010)

27. Rossi, R.A., Ahmed, N.K.: Social Networks — Network Data Repository [Datasets] (2015). http://networkrepository.com/

28. Newman, M.J.: A measure of betweenness centrality based on random walks. Soc. Netw. **27**(1), 39–54 (2005). https://doi.org/10.1016/j.socnet.2004.11.009

Adaptive Iterative Destruction Construction Heuristic for the Firefighters Timetabling Problem

Mohamed-Amine Ouberkouk(✉) ⓘ, Jean-Paul Boufflet ⓘ, and Aziz Mourkim ⓘ

Université de technologie de Compiègne, CNRS, Heudiasyc (Heuristics and Diagnosis of Complex Systems), CS 60 319 - 60 203, Compiègne Cedex, France
{mohamed-amine.ouberkouk,jean-paul.boufflet,aziz.moukrim}@hds.utc.fr

Abstract. Every year, wildfires accentuated by global warming, cause economic and ecological losses, and often, human casualties. Increasing operating capacity of firefighter crews is of importance to better face the forest fire period that yearly occurs. In this study, we investigate the real-world firefighters timetabling problem (FFTP) of the INFOCA institution in Andalusia (Spain) with the aim of increasing operating capacity while taking into account work regulation constraints. We propose an Integer Linear Programming model and an Adaptive Iterative Destruction Construction Heuristic solution approach to address the problem. We report on experiments performed on datasets generated using real-world data of the INFOCA institution. The work was initiated as part of the GEO-SAFE project (https://geosafe.lessonsonfire.eu/).

Keywords: Timetabling · Firefighters · ILP · Adaptive destruction/contruction heuristic

1 Introduction

Timetabling problems [1,7,9] involve allocating resources within time slots considering a predefined planning horizon while respecting precedence, duration, capacity, disjunctive and distribution (spacing, grouping) constraints. Staff planning aims at building timetables so that an organization can meet demands for goods or services. For each staff member, working and rest days are scheduled in a timetable while taking into account work regulation constraints and local regulation constraints, if any.

The first works on personnel scheduling can be traced back to Edie's work on traffic delays at toll booths [5]. Since then, scheduling algorithms have been applied in a lot of areas like transportation systems (airlines, railways), healthcare systems, emergency services (police, ambulances), call centers and other services (hotels, restaurants, commercial stores).

Comprehensive literature reviews covering a wide area of problems with many references on personnel scheduling can be found in [6,10]. The works are classified

B. Dorronsoro et al. (Eds.): META 2021, CCIS 1541, pp. 33–47, 2022.
https://doi.org/10.1007/978-3-030-94216-8_3

by type of problem, application area and solution method. As an example, the nurse rostering [4] is a scheduling issue in health systems. The objective is to build a daily schedule for nurses with the aim of obtaining a full timetable over few weeks for the institution. The rosters should provide suitably qualified nurses to cover the demand of working shifts arising from the numbers of patients in the wards. The resulting schedule should comply with regulatory constraints and should ensure that night and weekend shifts are fairly distributed while accommodating nurse preferences.

Staff scheduling is known as crew scheduling in transportation systems areas such as market/airlines, railways, mass transit and buses [2]. For these problems, there are two common features. The first is that both temporal and spatial constraints are involved. Each task is characterized by its starting time and location, and, its ending time and location. The second is that all tasks to be performed by employees are determined from a given timetable. The tasks are determined following a decomposition of the different duties that the company must ensure within a planning period. A task may be assuring a flight leg in airlines or ensuring a trip between two segments in a train.

The firefighters problem that we address consists in providing the INFOCA's daily schedule within a fixed planning horizon for a number of firefighter crews. Each firefighter is assigned to a crew for a year. These firefighters crews can be assigned to several types of shifts such as helicopter work, night work, work on demand (24 h on call). The planning period is the high-risk period from 1st June to 15th October where wildfires yearly occur (forest fire period).

The objective is to build a schedule for every crew of firefighters, hence a full timetable that covers all the forest fire period. The aim is to maximize the overall operating capacity while respecting the minimum demands for each shift, the regulatory constraints imposed by the institution as well as other soft constraints of good practice in order to make the schedules adequate to the preferences of the institution. The constraints of good practice relate to the grouping of assignments of same shifts within consecutive days, the allocation of compensations after rest days while maximizing of the number of operational crews a day.

The application of various metaheuristics to employee scheduling problems is presented in the reviews mentioned above. In this study, we choose to investigate an algorithm mainly based on an Adaptive Iterative Destruction/Construction Heuristic (AIDCH) [3]. An initial feasible solution that only complies with the minimum demands is build first by applying a constructive heuristic. Then, the AIDCH approach that we propose aims at increasing the overall operating capacity by first partly destroying a solution, next it is completed by inserting as many crews as possible, that can be easily done through a Destruction/Construction Heuristic approach. While completing the solution to increase the overall operational capacity, we make work together adaptive diversification mechanisms and parallel independent searches to avoid to be trapped in a local optimum.

In this paper we propose an Integer Linear Programming (ILP) formulation together with an Adaptative Iterative Destruction Construction Heuristic

(AIDCH) to address the firefighters timetabling problem (FFTP) of the INFOCA institution. The ILP is designed for modeling purposes and with the aim of giving lower bounds useful for the tuning analysis of the AIDCH solution approach. The Adaptive Iterative Destruction/Construction Heuristic is composed of an adaptive diversification mechanism at the destruction phase followed by an adaptive construction phase, based on a Best Insertion Algorithm, which performs parallel independent searches. The initial parameter values are adjusted by the algorithm according to the solution progress throughout the resolution process. The AIDCH is appropriate to generate solutions of good quality for the larger instances. The remainder of the paper is organized as follows. Section 2 provides a description of the FFTP, then the ILP formulation is presented in Sect. 3. The proposed AIDCH solution approach is described in Sect. 3. Computational experiments performed on a benchmark that we generated using real data of the INFOCA firefighter institution are reported in Sect. 4. Conclusion and future works are given in Sect. 5.

2   Problem Description

In this section we present a global overview of the real-world firefighter planning problem that we address. We gives the set of daily working shifts to be considered, we introduce the hard constraints to be respected and the soft constraints used to assess the quality of a solution.

The notations used for the types of shifts and their brief descriptions are the following:

(T12) from 8 am to 4 pm at fire station, regular daily shift;
(T16) from 3 pm to 10 pm at fire station, regular daily shift;
(H) from 8 am to 4 pm at fire station, regular daily shift, assigned to a helicopter;
(N) from 10 pm to 8 am at fire station, regular night shift;
(G7) from 7 am to 3 pm at fire station, stand-by to face instantly any extra urgent request;
(G24) 24 h guard, crew stay at home but may be mobilized to face any urgent situation;
(A3) from 8 am to 6 pm at fire station (or elsewhere) for training purposes;
(R) rest day;
(C) additional compensation day granted when a number of hours have been worked.

For the considered firefighters timetabling problem, the hard constraints relating to work regulation and to local regulation of the INFOCA institution are the following:

(H1) one shift a day: a firefighter crew can only be assigned to one shift a day;
(H2) minimum demands: each daily shift has a minimum demand of firefighter crews;

(H3) forbidden shift successions: some shift assignments on consecutive days are forbidden;

(H4) maximum workload: over the planning horizon, a maximum workload for every crew should not be exceeded;

(H5) compensation: compensation days are granted according to the hours worked, they should be used;

(H6) maximum consecutive working days: every firefighter crew have a maximum number of consecutive working days.

Some consecutive shift assignment are forbidden for a crew (H3), for instance a night shift ends at 8 am and cannot be followed by an helicopter shift which begins at 8 am, this forbidden consecutive shift assignment is denoted as (N, H).

Soft constraints are constraints of good practice that should be satisfied as best as possible. The violation of any soft constraint induces a penalty. A weighted sum of the penalties measures the quality of the solution produced. For the studied firefighters timetabling problem, the soft constraints are the following:

(S1) shift grouping: assignments of a crew to the same shift should be grouped. Each shift assignment change between two consecutive days is penalized;

(S2) same start time: start times should be the same whatever the working shifts over consecutive working days. Each starting time change for working shifts between two consecutive days is penalized;

(S3) compensation assignments: compensation day assignments should be right after the rest days, the aim is to allow firefighters to have a short vacation during the planning period. Each assignment of compensation not right after rest days is penalized.

(S4) period fairness: for the sake of fairness the workload should be balanced between the crews over the planning period. The unbalance of workload between crews should be minimized;

(S5) preferences: each crew assignment to an undesired shift is penalized;

(S6) evenly balance extra daily shifts: assigning of extra crews to the different shifts should be balanced each day. The unbalance on extra assignment to different shifts should be minimized each day.

Provided the minimum demand (H2) is respected, the idea beyond (S6) is to ensure a balance between shift assignments. If we can assign three extra crews for a day, we had better to assign a crew to three different shifts to balance operating capacity rather than assigning the three crews to a same shift.

3   ILP Model for FFTP

In this section we present the ILP model for minimizing the criteria detailed in Sect. 2. The ILP has a twofold objective, first a modeling purpose for investigating the problem we face, second we aim at obtaining optimal values whether possible for the smaller instances within a reasonable time limit (or lower/upper bounds). This allows to get reference values to make comparisons with the AIDCH solution approach that we propose. We present data and parameters prior to the decision variables, we then give the model.

The data and parameters are the following:

$Days$ set of days of the planning period, a day $d \in [1, \cdots, l_d]$, size n_d;

$Shifts$ set of types of shifts, a shift $s \in \{T12, T16, H, N, G7, G24, A3, R, C\}$, size n_s;

$Crews$ set of firefighter crews, size n_c;

l_d last day of the planning period;

F set of couples of forbidden consecutive shift assignment, e.g. $(N, H) \in F$;

r_s daily minimum demand for a working shift $s \in \{Shifts \setminus \{R, C\}\}$;

l_s duration of shift s (length in hours);

L maximum workload for any crew over the planning period;

t_s start time of shift s;

w_{oc} operating capacity weight;

w_{sg} shift grouping violation weight (S1);

w_{sst} same start time change violation weight (S2)

w_{ca} compensation assignments violation weight (S3);

w_p preferences violation weight (S5);

p_{csd} if crew c does not prefer to work on shift s on day d $p_{csd} = w_p$, zero otherwise (S5);

MAX_d maximum number of consecutive work days for a crew (H6);

WHC number of worked hours giving a compensation day.

The primary boolean variables are X_{csd}, if the crew c works on shift s in day d then $X_{csd} = 1$, zero otherwise. The secondary boolean variables used in the model are the followings:

$\alpha_{css'd} = 1$ if crew c works on shift s in day d and works on a different shift s' in day $d + 1$, zero otherwise;

$\beta_{css'd} = 1$ if crew c works on shift s in day d and works on a different shift s' in day $d + 1$ with $t_s \neq t_{s'}$, zero otherwise;

$\gamma_{css'd} = 1$ if the crew c works on shift s in day d with $s \neq' R'$ and is assigned to shift $s' =' C'$ in day $d + 1$, zero otherwise.

$\alpha_{css'd} = 1$ if a shift change violation occurs (S1, shift grouping), $\beta_{css'd} = 1$ if a working time change violation occurs (S2, same start time) and $\gamma_{css'd} = 1$ if a compensation assignment violation occurs (S3, compensation assignment).

The integer variables used in the model are the followings:

λ_d daily difference between the maximum number of assignable crews (n_c) and those assigned;

δ_c total number of worked shifts for crew c over the planning period;

θ_c total working time of crew c over the planning period;

ρ_{cd} number of worked hours of crew c from the first day to day d;

$\phi_{cc'}$ number of shift assignment difference between the crews c and c' (S4);

$\varphi_{cc'}$ working time difference between the crews c and c' (S4);

$\psi_{ss'}$ unbalance of assignments between the shifts s and s' (S6).

The aim is to maximize operating capacity over the planning period while minimizing the soft constraint violations. We propose the following ILP to address this problem:

Min

$$w_{oc} \cdot \sum_{d \in Days} \lambda_d \tag{1a}$$

$$\sum_{c \in Crews} \sum_{s \in Shifts \backslash \{R,\, C\}} \sum_{s' \in Shifts \backslash \{R,\, C\}} \sum_{d \in Days} (w_{sg} \cdot \alpha_{css'd} + w_{sst} \cdot \beta_{css'd} + w_{ca} \cdot \gamma_{css'd}) \tag{1b}$$

$$+ \sum_{c \in Crews} \sum_{c' \in Crews} (\phi_{cc'} + \varphi_{cc'}) \tag{1c}$$

$$+ \sum_{c \in Crews} \sum_{s \in Shifts \backslash \{R,\, C\}} \sum_{d \in Days} p_{csd} \cdot X_{csd} \tag{1d}$$

$$\sum_{s \in Shifts \backslash \{R,\, C\}} \sum_{s' \in S \backslash \{R,\, C\}} \psi_{ss'} \tag{1e}$$

Subject to:

$$\sum_{s \in Shifts} X_{csd} = 1 \quad \forall c \in Crews, \forall d \in Days \tag{2}$$

$$\sum_{c \in Crews} X_{csd} \geq r_s \quad \forall d \in Days, \forall s \in \{Shifts \backslash \{R,\, C\}\} \tag{3}$$

$$X_{csd} + X_{cs'(d+1)} \leq 1 \quad \forall (s, s') \in F, \forall c \in Crews, \forall d \in Days \backslash \{l_d\} \tag{4}$$

$$\sum_{s \in \{Shifts \backslash \{R,\, C\}\}} \sum_{d \in Days} l_s \cdot X_{csd} \leq L \quad \forall c \in Crews \tag{5}$$

$$\sum_{s \in \{Shifts \backslash \{R,\, C\}\}} \sum_{d' \in Days,\, d' \leq d} l_s \cdot X_{csd} = \rho_{cd} \quad \forall c \in Crews, \forall d \in Days \tag{6}$$

$$\sum_{d' \in Days,\, d' \leq d} X_{csd} \leq \frac{\rho_{cd}}{WHC} \quad s = \text{'C'}, \forall c \in Crews, \forall d \in Days \tag{7}$$

$$\sum_{d \in Days} X_{csd} = \left\lfloor \frac{\rho_{c(l_d)}}{WHC} \right\rfloor + 1 \quad s = \text{'C'}, \forall c \in Crews \tag{8}$$

$$\sum_{s \in \{Shifts \backslash \{R,\, C\}\}} \sum_{d' \leq (1+MAX_d),\, (d+d') \leq l_d} X_{csd} \leq MAX_d \quad \forall c \in Crews, \forall d \in Days \tag{9}$$

$$\sum_{c \in Crews} \sum_{s \in \{Shifts \backslash \{R,\, C\}\}} X_{csd} = n_c - \lambda_d \quad \forall d \in Days \tag{10}$$

$$X_{csd} + X_{cs'(d+1)} \leq 1 + \alpha_{css'd} \quad \begin{cases} \forall s, s' \in \{Shifts \backslash \{R,\, C\}\}, s \neq s' \\ \forall c \in Crews, \forall d \in \{Days \backslash \{l_d\}\} \end{cases} \tag{11}$$

$$X_{csd} + X_{cs'(d+1)} \leq 1 + \beta_{css'd} \quad \begin{cases} \forall s, s' \in \{Shifts \setminus \{R, C\}\}, \; s \neq s', \; with \; t_s \neq t_{s'} \\ \forall c \in Crews, \forall d \in \{Days \setminus \{l_d\}\} \end{cases}$$

$$\text{(12)}$$

$$X_{csd} + X_{cs'd+1} \leq 1 + \gamma'_{css'd} \quad \begin{cases} s \in \{Shifts \setminus \{R, C\}\}, \; s' = 'C' \\ \forall c \in Crews, \; \forall d \in \{D \setminus \{l_d\}\} \end{cases} \quad \text{(13)}$$

$$\sum_{s \in Shifts \setminus \{R, C\}} \sum_{d \in Days} X_{csd} = \delta_c \quad \forall c \in Crews \tag{14}$$

$$\sum_{s \in Shifts \setminus \{R, C\}} \sum_{d \in Days} l_s \cdot X_{csd} = \theta_c \quad \forall c \in Crews \tag{15}$$

$$\delta_c - \delta_{c'} \leq \phi_{cc'} \quad \forall c, c' \in Crews, \; c \neq c' \qquad \theta_c - \theta_{c'} \leq \varphi_{cc'} \quad \forall c, c' \in Crews, \; c \neq c'$$

$$\text{(16)} \qquad\qquad\qquad\qquad\qquad\qquad \text{(17)}$$

$$\left(\sum_{c \in Crews} X_{csd} - r_s \right) - \left(\sum_{c \in Crews} X_{cs'd} - r_{s'} \right) \leq \psi_{ss'} \quad \begin{cases} \forall s, s' \in \{Shifts \setminus \{R, C\}\} \\ \forall d \in Days \end{cases}$$

$$\text{(18)}$$

$$X_{csd}, \; \alpha_{css'd}, \; \beta_{css'd}, \; \gamma_{css'd} \in \{0, 1\} \tag{19}$$

$$\delta_c, \theta_c, \rho_{cd}, \phi_{cc'}, \varphi_{cc'}, \psi_{ss'} \in \mathbb{N} \tag{20}$$

The five terms of the objective function aims at maximizing operating capacity while minimizing the soft constraint violations. The first term (1a) aims at maximizing operating capacity. The weighted sum (1b) assesses the (S1, shift grouping), (S2, same start time) and (S3, compensation assignments) soft constraint violations. The period fairness (S4) soft constraint relates to the number of shift assignment differences and to the working time differences between crews, they are considered using the (1c) term. The preferences of the firefighters (S5) are considered using the (1d) term. The evenly balance of extra daily shifts (S6) is considered using the (1e) term.

The hard constraints **one shift a day** (H1) are enforced by Eq. (2). The hard constraints **minimum demands** (H2) are enforced by Eq. (3). The hard constraints **forbidden shift successions** (H3) are enforced by Eq. (4). The hard constraints **maximum workload** (H4) are enforced by Eq. (5). The hard constraints **compensation** (H5) are enforced by Eqs. (6)–(8). For a crew c and a day d, Eq. (6) count ρ_{cd}, the number of worked hours of crew c from the first day of the planning period to day d, and links variables X_{csd} and ρ_{cd}. For a crew c and a day d, Eq. (7) forces the number of compensation days ($s =' C'$) being assigned to be less or equal to (ρ_{cd}/WHC) since one compensation day is granted when WHC worked hours are made. For a crew c, all the compensation days must be assigned over the planning horizon (until $d = l_d$), this is enforced by Eq. (8). The hard constraints **maximum consecutive working days** (H6) are enforced by Eq. (9). For a crew c and a day d, the crew is assigned to at most MAX_d consecutive working shifts (rest and compensation days are not to be considered).

The daily differences between the maximum number of assignable crews (n_c) and those assigned are to be minimized to optimize the overall operating capacity, the λ_d values are assessed by Eq. (10).

Consider a crew c, two days d and $d + 1$, if the crew is assigned to two different shifts ($s \neq s'$) a **shift grouping** (S1) soft constraint violation occurs and Eq. (11) sets $\alpha_{css'd} = 1$. Consider a crew c, two days d and $d + 1$, if the crew is assigned to two different shifts ($s \neq s'$) and the start times of these shifts are different ($t_s \neq t_{s'}$) a **same start time** (S2) soft constraint violation occurs and Eq. (12) sets $\beta_{css'd} = 1$. Consider a crew c, two day d and $d + 1$, if the crew is assigned to a working shift ($s \neq$ 'R') on day d, and if this crew is assigned to a compensation day ($s' = $ 'C') on day $d + 1$ a **compensation assignment** (S3) soft constraint violation occurs and Eq. (13) sets $\gamma_{css'd} = 1$. Every compensation day assignment will be right after a rest day (constraints of good practice imposed by the institution).

Consider a crew c, Eq. (14) counts δ_c the total number of worked shifts over the planning period and Eq. (15) counts θ_c the total working time over the planning period. Hence, Eq. (16) gives $\phi_{cc'}$ the number of shift assignment differences. Given that $\phi_{cc'} \in \mathbb{N}$, a negative difference involves $\phi_{cc'} = 0$, so for any couple of crews only positive differences are counted. The same rationale applies on Eq. (17) for $\varphi_{cc'}$, the number of working time differences. These variables $\phi_{cc'}$ and $\varphi_{cc'}$ are used for the **period fairness** (S4) soft constraint violations assessment.

We recall that **preferences** (S5) soft constraint violations are assessed by Eq. (1d).

Consider a day d and two shifts s and s', Eq. (18) aims at **evenly balance extra daily shifts** (S6). Minimum demands (H2) are enforced by Eq. (3), assigning of extra crews to shifts should be balanced each day within the forest fire period to increase operating capacity.

Equation (19) defines variables X_{csd}, $\alpha_{css'd}$, $\beta_{css'd}$ and $\gamma_{css'd}$ as boolean. Equation (20) defines variables δ_c, θ_c, ρ_{cd}, $\phi_{cc'}$, $\varphi_{cc'}$ and $\psi_{ss'}$ as integers.

4 Adaptive Iterative Destruction/Construction Heuristic

We propose an Adaptive Iterative Destruction/Construction Heuristic (AIDCH) to compute solutions of good quality for larger instances of the FFTP. The Algorithm 1 gives the global scheme of the AIDCH proposed approach. We use the adaptive construction approach *BuildFeasibleSchedule()* to build an initial solution which respects the hard constraints. The initial solution complies with *minimum demands* (H2) but there is room for improvement in operating capacity.

Algorithm 1: General structure of AIDCH

Input	: An instance of FFTP
Output	: S_{best} best solution found
Parameters	: D_{limit} limit for diversification degree, n_c number of crews
	n_s number of type of shifts
Variables	: $iter$ number of iterations, $MaxIter$ maximum iteration
	D_{max} diversification degree, S_{cur} current solution

$iter := 0$
$MaxIter := n_c$
$D_{max} := 3$
$D_{limit} := \left\lceil \frac{n_c}{n_s} \right\rceil$
$S_{cur} := $ BuildFeasibleSchedule()
$S_{best} := S_{cur}$
while $iter < MaxIter$ **do**
    $k := $ rand$(1, D_{max})$
    AdaptiveDestruction(S_{cur}, k) /* adaptive diversification */
    AdaptiveConstruction(S_{cur}) /* insert as many crews as possible in S_{cur} */
    if $S_{cur} > S_{best}$ **then**
        $S_{best} := S_{cur}$
        $iter := 0$
        $D_{max} := 3$
    else
        $iter + +$
        $D_{max} := \min(D_{max}+1, D_{limit})$
    end
end

Provided a feasible solution, at each iteration, a part of the solution is destroyed by removing at random a number k of crews, then it is completed by inserting as many crews as possible in order to increase the operating capacity (while respecting the hard constraints). At each overall iteration at most D_{max} crews are removed ($k \leq D_{max}$). Therefore, we define D_{max} as the degree of diversification. The D_{max} value is initialized to 3, next incremented after each non-improving overall iteration up to D_{limit}. We set $D_{limit} = \lceil n_c/n_s \rceil$ which represents the average number of crews that can be assigned to shifts. Provided an improvement is found, D_{max} is reset to 3 to entirely explore the neighborhood of the new solution. We perform an adaptive construction procedure to complete the solution. This process is reiterated and it stops when $MaxIter$ overall iterations have been performed without improving the quality of the solution. We set $MaxIter = n_c$. The final result is the best solution found over all iterations.

Algorithm 2: Best Insertion Algorithm

> **Input**   : S_{cur} a partial solution
> $(\alpha, \beta, \gamma, \theta, \omega, \mu)$ parameter set
> **Output** : S_{best} best solution found
> **Variables** : $(d,s,c)^*$ best triplet, success boolean
> $S_{best} := S_{cur}$ /* store reference solution for BIA */
> success := true
> **while** *success* **do**
> $(d,s,c)^* := (\emptyset, \emptyset, \emptyset)$
>   **foreach** $d \in Days$ **do**
>   **foreach** $s \in Shifts$ **do**
>   **foreach** $c \in Crews$ **do**
>   ComputeBIC(d,s,c)
>   UpdateBestTriplet (d,s,c)*
>   **end**
>   **end**
>   **end**
>   success := Insert(S_{cur}, (d,s,c)*) /* if no feasible insertion, Insert returns false */
>   /* Comparing S_{cur} and S_{best}, all terms of the objective function are assessed */
>   **if** $S_{cur} > S_{best}$ **then**
> $S_{best} := S_{cur}$
>   **end**
> **end**

The proposed AIDCH algorithm makes use of an adaptive diversification mechanism with the aim to escape from local optima. We explore the neighborhood of the new solution as soon as an improvement is found. We explore more distant zones by increasing D_{max} whenever the search is trapped in a local optimum.

The main component of the AIDCH heuristic is the *AdaptativeConstruction(S_{cur})* procedure, an adaptive construction heuristic based on a Best Insertion Algorithm (BIA) shown in Algorithm 2. The BIA algorithm considers a partial solution S_{cur}, and tries to insert as many crews as possible in S_{cur}, one by one. At each iteration, the BIA assesses all feasible insertions that respect the hard constraints and scores them according to a Best Insertion Criterion (BIC). The best insertion is then performed and the quality of S_{cur} is assessed considering all terms of the objective function (1a)–(1e). This process is iterated until no more valid insertion is possible. The algorithm returns the updated S_{cur}, the best solution over all the BIA iterations.

To evaluate the insertion of a crew in the planning (day, shift), we propose to compute the Best Insertion Criterion (BIC) as follows:

$$(SG^\alpha * SST^\beta * CA^\gamma * PF^\theta * P^\omega * EB^\mu)$$

The aim is to minimize the soft constraints violation whether the insertion is performed. In case a hard constraint is violated (e.g. maximum workload (H4)), the BIC is set to $+\infty$. The criterion is composed of 6 terms, one for each soft constraints: SG is for the **S**hift **G**rouping (S1), SST is for the **S**ame **S**tart **T**ime (S2), CA is for the **C**ompensation **A**ssignments (S3), PF is for the **P**eriod **F**airness (S4), P is for the **P**references (S5) and EB is for the **E**venly **B**alance extra daily shifts (S6). The terms are weighted with parameters α, β, γ, θ, ω and μ in order to control their relative importance.

At each iteration i of the AIDCH heuristic, $AdaptativeConstruction(S_{cur})$ works as follows. Four constructive heuristics launch separately BIA with different values of the parameter set $(\alpha, \beta, \gamma, \theta, \omega, \mu)$ on the current solution. During each launch, α, β, γ, θ, ω and μ are chosen randomly in the 6 dimension space having the center $(\alpha_{i-1}, \beta_{i-1}, \gamma_{i-1}, \theta_{i-1}, \omega_{i-1}, \mu_{i-1})$ and the side length ϕ, where α_{i-1}, β_{i-1}, γ_{i-1}, θ_{i-1}, ω_{i-1} and μ_{i-1} are the best parameters obtained by the method at previous iteration. All four BIA being applied, the parameter set that produces the best solution is stored to be used in the next iteration.

Finally, the best solution obtained among the four methods is retained as the current solution. This aims at performing parallel independent searches in the solutions space and at choosing the best values of the parameters to better explore the solutions space to speed-up the convergence of the AIDCH algorithm toward a good solution.

5   Computational Experiments

In our experiments, our objectives were: (i) to show the adaptive construction impact, by comparing ϕ together with the best parameter set that produces the best solution at previous iteration to compute the next parameter set, versus a fully randomized parameter set; (ii) to show the efficiency of the adaptive destruction, impact of an adaptive D_{max} for perturbations versus a constant one; (iii) to compare performances between the ILP model and the AIDCH approach within a 3600 s time limit.

Tests were done using C++ compiled with gcc version 7.5.0, using STL, using a CPLEX 12.10 [8] solver with a single thread and the *MipEmphasis* parameter set to *feasibility*, on a machine with an Intel(R) Xeon(R) X7542 CPU @ 2.6 GHz and 64 GB of RAM.

Datasets Overview and Performance Metric
We tested the ILP and AIDCH approaches on a benchmark composed of 4 datasets, each having 7 instances, that we generated using real data of the INFOCA firefighter institution. Datasets have been created to be of increasing difficulty, the firsts of reasonable sizes given that the ILP may face difficulty to get a solution within the time limit. The instances in datasets are ranged according to the number of crews n_c and to the total daily number of working shifts demands (i.e. $\sum r_s$). So, instances are denoted as $cXXrYY(a/b)$, the (a/b) notation is used whether n_c and $\sum r_s$ equals for two distinct instances which are different in minimum demands distributions.

For each instance, the AIDCH algorithm is run 10 times. We recorded the **R**elative **P**ercentage **E**rror, we defined as $RPE = 100 * (Z_{best} - Z_{max})/Z_{best}$ and the **A**verage **R**elative **P**ercentage **E**rror, we defined as $ARPE = 100 * (Z_{best} - Z_{avg})/Z_{best}$ where Z_{max} is the best result obtained among the ten executions, Z_{avg} is the average result obtained among the ten runs and Z_{best} is the best solution found by the AIDCH approach for the according instance. The ARPE criterion aims at investigating whether the AIDCH is stable over the runs.

To compare the solutions found by the AIDCH approach against the solutions attained by the ILP approach, we define the **R**elative **P**ercentage **G**ap as $RPG = 100 * (Z_{ILP} - Z_{max})/Z_{ILP}$ where Z_{ILP} represents the solution value attained, if any, by the ILP approach for an instance.

For our experiments using the ILP, we set w_{oc} to 2, w_{sg} to 1, w_{sst} to 1, w_{ca} to 1 and w_p to 2.

Impact of the Adaptive Construction Mechanism

We first carried out preliminary experiments to choose the best value of ϕ that is necessary to show the impact of the adaptive construction mechanism, because of lack of space those experiments are not reported here. According to these experiments, the parameter value $\phi = 0.1$ provides the best results considering RPE.

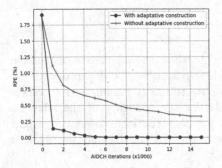
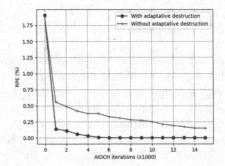

Fig. 1. Adaptive construction impact   **Fig. 2.** Adaptive destruction impact

The adaptive construction mechanism aims to guide the search by computing at each time the best trade-off between the different terms of the BIC representing soft constraints violations. To show whether it is efficient, we conducted experiments with the adaptive construction mechanism and without the adaptive construction mechanism. In that latter case, the parameters of BIC are chosen randomly in $[0, 1]$ at each iteration. In these experiments, for each instance, the algorithm is launched and we record the best solution for the first 15000 iterations. We performed these tests using 2 instances chosen at random from each dataset. We report in Fig. 1 the average of RPE values computed for the 8 chosen instances against the number of iterations.

As it can be shown in Fig. 1, the adaptive construction mechanism permits to converge faster toward good solutions rather than without adaptive construction mechanism.

Impact of the Diversification Mechanism

To evaluate the effectiveness of the adaptive destruction, we tested a version of AIDCH where the diversification degree D_{max} is set to 3. As aforementioned, we record the best solution for the first 15000 iterations using this fixed value. We proceed in the same way using the adaptive diversification mechanism that makes use of D_{max} to explore the neighborhood of the new solution as soon as an improvement is found and also to explore more distant zones whenever the search is trapped in a local optimum.

Figure 2 shows the average of RPE values recorded against the number of iterations for these two versions. The adaptive diversification mechanism, achieved using the management of D_{max}, permits to converge faster toward good solutions rather than without its use.

Based on these two graphs, we can easily notice that the average of RPE values with the adaptive mechanisms is always below the average of RPE values with the standard perturbation at each iteration, which shows the effectiveness of our proposed technique.

Table 1. Performances of ILP and AIDCH approaches

Instance	ILP	t (s)	gap	AIDCH	t (s)	RPG	ARPE	Instance	ILP	t (s)	gap	AIDCH	t (s)	RPG	ARPE
c18r09a	1325	1443	0	1325	341	0	0	c50r22a	ns	-	nc	3765	741	nc	0.43
c18r10a	1359	1409	0	1359	352	0	0	c50r23a	ns	-	nc	3783	754	nc	0.31
c18r10b	1344	1526	0	1344	348	0	0	c50r26a	ns	-	nc	3799	759	nc	0.27
c18r11a	1378	1886	0	1378	372	0	0	c50r28a	ns	-	nc	3823	783	nc	0.58
c18r11b	1420	2786	0	1420	401	0	0	c50r31a	ns	-	nc	3947	849	nc	0.52
c18r12a	1422	2103	0	1422	391	0	0	c50r33a	ns	-	nc	3931	817	nc	0.56
c18r12b	1440	2209	0	1440	413	0	0	c50r35a	ns	-	nc	4097	831	nc	0.34
c30r15a	1767	-	0.74	1758	553	−0.51	0.1	c70r31a	ns	-	nc	4913	943	nc	0.67
c30r16a	1801	-	1.18	1811	561	0.56	0.15	c70r33a	ns	-	nc	4957	954	nc	0.71
c30r17a	1818	-	0.44	1860	582	2.31	0.22	c70r37a	ns	-	nc	5102	995	nc	0.69
c30r18a	1834	-	0.11	1867	593	1.80	0.08	c70r40a	ns	-	nc	5151	1034	nc	0.65
c30r19a	1889	-	0.48	1934	612	2.38	0.13	c70r44a	ns	-	nc	5213	1067	nc	0.71
c30r20a	2144	-	12.9	1947	661	−9.19	0.26	c70r47a	ns	-	nc	5557	1113	nc	0.83
c30r21a	1966	-	2.77	1936	657	−1.53	0.16	c70r50a	ns	-	nc	5401	1158	nc	0.67

ILP Versus AIDCH

Table 1 compares the results obtained by the ILP solver against those obtained by the AIDCH approach. In Table 1, **ns** stands for no solution, **nc** stands for not calculable, and - shows that the 3600 s time limit has been attainted. For the sake of compactness, datasets are grouped by two then tabulated side by side. Column *Instance* gives the instance label. The next tree columns, *ILP, t (s)* and *gap* show the performances of the ILP. They report the objective function value, the computing time and the gap found by the CPLEX solver. Then, the next four columns, *AIDCH, t (s)*, RPG, and *ARPE* show the performances of the AIDCH approach. They report the objective function value, the computing time, the gap between the solutions found by the AIDCH approach and the solution provided by the ILP solver and the average of RPEs over the 10 runs for an instance.

The ILP approach attains optimal solutions for all $n_c = 18$ instances. It faces difficulty for the second dataset having $n_c = 30$, however feasible solutions are obtained within the 3600 s time limit. For the third and the fourth datasets having $n_c = 50$ and $n_c = 70$, the ILP approach fails to find a feasible solution within the time limit.

For the first dataset, the AIDCH approach succeeded in obtaining all the optimal solutions found by the ILP approach. We also notice that all the ARPE values are equal to 0: which means that the AIDCH approach was able to attain the optimal solutions.

For the second dataset, the AIDCH approach attains solutions closed to or better than the solutions obtained by the ILP approach within a 3600 s time limit. For four instances the RPG values are between 0.56 and 2.38. For the three other instances, the AIDCH approach obtains better solutions than the ones provided by the ILP approach, with an RPG values from −0.51 up to −9.19. ARPE values are less than 0.26 for all instances which shows the stability of our proposed heuristic approach for this dataset.

For the third and the fourth datasets, the AIDCH approach was able to find solutions in a reasonable time. The ARPE values are less than 0.83, the proposed heuristic behaviour is stable over the last two datasets. Unfortunately, the quality of the solutions found by the AIDCH approach cannot be assessed since the ILP approach fails to provide solutions for these datasets within the one hour time limit.

6   Conclusion and Future Work

We presented in this paper both an ILP model and a AIDCH heuristic to address the real-worl firefighters timetabling problem (FFTP) of the INFOCA institution. The proposed approaches were tested over four datasets with different sizes of increasing difficulty that we generated using real data from INFOCA. The ILP approach obtained optimal or near optimal solutions for the first two datasets, but it faced difficulty in obtaining feasible solutions for the larger instances of the two other datasets. The AIDCH approach obtained good solutions for all the instances of the first two datasets, those are either optimal or closed to the ones obtained by the ILP approach. The proposed heuristic approach was able to find feasible solutions for the larger instances within a reasonable computation time. Future works aim at investigating a metaheuristic solution approach to improve the quality of the solutions obtained over the datasets and aim at reducing the computation time. We also plan to obtain lower bounds for the larger instances for comparison purposes.

Acknowledgments. This work was initiated as part of the GEO-SAFE project. The GEO-SAFE project has received funding from the European Union's Horizon 2020 research and innovation program under the Marie Skłodowska-Curie RISE grant agreement No 691161. This work is carried out in the framework of the Labex MS2T, which was funded by the French Government, through the program "Investments for the future" managed by the National Agency for Research (Reference ANR-11-IDEX-0004-02).

References

1. Aggarwal, S.C.: A focussed review of scheduling in services. Eur. J. Oper. Res. **9**(2), 114–121 (1982)
2. Barnhart, C., Cohn, A.M., Johnson, E.L., Klabjan, D., Nemhauser, G.L., Vance, P.H.: Airline crew scheduling. In: Hall, R.W. (ed.) Handbook of Transportation Science, pp. 517–560. Springer, Boston (2003). https://doi.org/10.1007/0-306-48058-1_14
3. Ben-Said, A., El-Hajj, R., Moukrim, A.: An adaptive heuristic for the capacitated team orienteering problem. In: 8th IFAC Conference on Manufacturing Modelling, Management and Control, pp. 1662–1666 (2016)
4. Burke, E.K., De Causmaecker, P., Berghe, G.V., Van Landeghem, H.: The state of the art of nurse rostering. J. Sched. **7**(6), 441–499 (2004)
5. Edie, L.C.: Traffic delays at toll booths. J. Oper. Res. Soc. Am. **2**(2), 107–138 (1954)
6. Ernst, A.T., Jiang, H., Krishnamoorthy, M., Owens, B., Sier, D.: An annotated bibliography of personnel scheduling and rostering. Ann. Oper. Res. **127**(1), 21–144 (2004)
7. Ernst, A.T., Jiang, H., Krishnamoorthy, M., Sier, D.: Staff scheduling and rostering: a review of applications, methods and models. Eur. J. Oper. Res. **153**(1), 3–27 (2004)
8. IBM, Cplex, User's Manual (2020). https://www.ibm.com
9. Tien, J.M., Kamiyama, A.: On manpower scheduling algorithms. SIAM Rev. **24**(3), 275–287 (1982)
10. Van den Bergh, J., Beliën, J., De Bruecker, P., Demeulemeester, E., De Boeck, L.: Personnel scheduling: a literature review. Eur. J. Oper. Res. **226**(3), 367–385 (2013)

Automatic Generation of Metaheuristic Algorithms

Sergio Iturra[1], Carlos Contreras-Bolton[2](✉) [iD], and Victor Parada[1,3] [iD]

[1] Departamento de Ingeniería Informática, Universidad de Santiago de Chile,
Santiago, Chile
{sergio.iturra,victor.parada}@usach.cl
[2] Departamento de Ingeniería Industrial, Universidad de Concepción,
Concepción, Chile
carlos.contreras.b@udec.cl
[3] Instituto Sistemas Complejos de Ingeniería (ISCI), Santiago, Chile

Abstract. Designing a heuristic algorithm to solve an optimization problem can also be seen as an optimization problem. Such a problem seeks to determine the best algorithm contained in the search space. The objective function corresponds to the computational performance of the algorithm measured in terms of computational time, complexity, number of instructions or number of elementary operations. The automatic design of algorithms has been explored for several combinatorial optimization problems. In this work, we extend this exploration towards the automatic design of metaheuristics to find solutions for the traveling salesman problem. The process is carried out by genetic programming. The resulting algorithms are combinations of well-known metaheuristics and, in some cases, present better computational performance than the existing algorithms for the set of selected test instances.

Keywords: Automatic generation of algorithm · Genetic programming · Metaheuristics · Traveling salesman problem

1 Introduction

There is a family of optimization problems that come from various fields of knowledge and are characterized by the tremendous computational difficulty that arises when trying to determine an optimal solution. Such optimization problems, which we call complex problems here, are considered difficult because currently, a polynomial and exact algorithm that can solve all the instances of a problem with computational efficiency is not known [15,25]. In this field, it is accepted that an algorithm is efficient when it requires a number of steps that grow polynomially with the input. A typical strategy for addressing complex optimization problems is through mathematical programming, which considers an objective function that corresponds to the criteria to be optimized and a set of constraints that define the solution space that contains the optimal solution. Integer programming algorithms utilize the enumeration of the solution space, a

© Springer Nature Switzerland AG 2022
B. Dorronsoro et al. (Eds.): META 2021, CCIS 1541, pp. 48–58, 2022.
https://doi.org/10.1007/978-3-030-94216-8_4

task that may require very high computational time or memory even when dealing with small instances of the problem [6, 32]. This challenge constitutes one of the main fields of scientific research in the area of combinatorial optimization, and it is a relentless search to improve existing techniques or to find new methods for addressing this situation. The motivation behind this search consists of innumerable practical situations that occur in various areas of knowledge, such as transport [39], health care [14], sports [41], production processes [34], and logistics [40].

One of the most commonly used practical approaches for addressing the family of complex optimization problems considers metaheuristics. A metaheuristic is a method that describes a general procedure to effectively inspect the solution space of an optimization problem and thus determine the best solution inspected [36]. In recent decades, this field has increased considerably because numerous metaheuristics have been generated, and a wide variety of problems have been studied under this approach [11]. Although a metaheuristic does not guarantee the determination of the optimal solution, in practice, they are very effective because they require a low computational time and provide a solution close to the optimal, and in many instances, they return the optimal solution. Such techniques have originated analogizing with different phenomena in nature, such as the species evolution, particle swarms, bee and ant colonies, and pure substance cooling. In general, they can be classified into single-solution search, population-based search, or hybrid metaheuristics [36]. Although their origins are varied, some characteristics are shared by several metaheuristics: a) they carry out the search process by gradually visiting solutions that belong to the problem-space, b) they work on a current solution or a current set of solutions in every step, c) the problem optimization function inherently guides the search process, d) they use exploration and exploitation strategies and e) they partially store the search space.

Recent literature has shown the emergence of a wide variety of hybrid metaheuristics that have better computational performance than the same metaheuristics used individually [4, 5, 13, 29, 38]. Such hybrid algorithms arise when considering the best components of metaheuristics and assembling them appropriately for the optimization problem at hand. The variety of works considers hybridizations of metaheuristics, with other metaheuristics, constraint programming, search tree techniques, and mathematical programming. However, the main deficiency that arises in this field is the design step because it is difficult to know in advance the appropriate combination for each optimization problem. It is necessary to identify how many and what components can be integrated to generate the hybrid algorithm that responds with good computational performance for the specific optimization problem. A standard or practical guide that facilitates the design task is not established in the literature. In practice, the authors find the appropriate combination of components through computational experiments and manually test some possibilities among the many possible combinations. Nor is there a standard method for carrying out experimentation, and although today there is a technological advance that allows a large number of numerical tests, to the best of our knowledge, the automation of this process has not been explored.

Another approach used to address complex optimization problems is the automatic generation of an algorithm (AGA). The AGA automatically assembles elementary components that potentially compose an algorithm for a given optimization problem [7,8,31,33]. This task is possible because determining the best algorithm for an optimization problem is also a master optimization problem. Consequently, the search for the best algorithm for a given problem reduces to solving the master problem by some of the existing methods, which in practice can be any of the current metaheuristics. The elementary components that can be considered are diverse; they can be specific heuristics already existing for the problem, the atomic parts of such heuristics, or exact algorithms of mathematical programming. Genetic programming (GP) is particularly appropriate for this task because it artificially evolves populations of syntactic trees that represent combinations of instructions, such as those that occur in an algorithm [27]. In this way, several algorithmic combinations can be represented with syntax trees and combined by evolutionary computing. This technique has allowed the generation of new algorithms for combinatorial optimization problems [3,9,19,26], a fact that suggests that the same technique could automatically produce hybrid metaheuristics. AGA is not only an automatic method for combining thousands of components and exploring the space composed of all hybrid metaheuristics but also determining the appropriate algorithm for each optimization problem, thus providing an experimental standard for this field of knowledge.

In this work, we use AGA to generate single-solution hybrid metaheuristic algorithms for the traveling salesman problem (TSP). The algorithms are constructed through GP by evolving syntactic trees [1]. The components of syntactic trees are functions and terminals, which are instructions typically used to write pseudocode and primary components typically considered in the heuristic, metaheuristic and exact methods. In addition, a set of instances is selected and divided into two groups: the first group is used for the construction of metaheuristics, and the second group is used to evaluate the already constructed hybrid metaheuristics.

In the following section, the literature review is presented. The procedures for generating the metaheuristic algorithms are described in the third section. The computational results of the generated algorithms are presented in the fourth section. The conclusions of the study are presented in the last section.

2   Literature Review

Only recently have the first attempts to automatize the design of hybrid metaheuristics appeared. One of them is the novel approach by Hassan and Pillay [12] that proposes a meta-GA to automate the hybridization of metaheuristics. The authors consider the following algorithms: simulated annealing, tabu search, iterated local search and memetic algorithm to solve the TSP. Additionally, Hassan and Pillay [12,13] efficiently solved the aircraft landing problem and the two-dimensional bin packing problem using the same approach. Their auto-mated designed hybrid metaheuristics showed better performance than individ-ually used metaheuristics and manually created hybrid metaheuristics. Recently, Tezel and Mert [37] propose a work that uses fuzzy logic and fuzzy systems to create a cooperative scheme for the automatic selection of proper metaheuris-tic algorithms and control searching process dynamically. Their approach was tested on 0–1 knapsack problem, and the computational experiments showed to be much more effective in searching the solution space, although, it did not dif-fer from the other algorithms in terms of computing times. Although these first attempts to automatize the process in the literature have generated promising results, some drawbacks have been observed, such as depending on predefined templates (structures), only selection automatically of sequences of metaheuris-tics to use, and the resulting algorithms with their analysis are not shown. The first implies that several successful combinations within the search space may not be visited due to fixed structures. The second means that new automati-cally generated algorithms are only sequences of metaheuristics that are run one after another. Therefore, there is no hybridization among components of a meta-heuristic with another metaheuristic completely different. The third means that these new automatically generated algorithms are not available to the scientific community in the field.

From the research line of automatic algorithm configuration [35], attempts are also being made to generate hybrid metaheuristics using tools that are typ-ically used in this area, such as IRACE [16]. Thus, in [17], instead of seek-ing parameters, they sought metaheuristic components and tuned the param-eters, all in a single process. The authors used a predefined framework, with grammars for each metaheuristic to generate hybrid metaheuristics for three combinatorial optimization problems. Other similar approaches were previously carried out by Marmion et al. [21], who used tools to design stochastic local searches automatically and non-hybrid metaheuristics as ACO algorithms [18]. Alfaro-Fernández et al. [2] generated hybrid metaheuristics following the pre-vious methodology described for solving hybrid flowshop scheduling problems. Their algorithms are competitive against state-of-the-art algorithms. Recently, Pagnozzi and Stützle [22–24] proposed an automatic design system of stochas-tic local search for permutation flowshop problem and other two variants that consider additional constraints. The approach uses a configuration tool to com-bine algorithmic components following a set of rules defined as a context-free grammar. Their experiments show that the generated algorithms outperform the state-of-the-art. However, these approaches from the configuration of algo-rithm parameters have some limitations, such as the use of predefined templates and grammars that limit the search space. The authors did not use a specific search method for the new task of searching for potential new combinations, but instead, they used IRACE, which is a specialized method for tuning parameters; thus, they possibly failed to explore possible good combinations. Besides, some approaches are limited to only some metaheuristics, missing the opportunity to mix, for instance, metaheuristics based on a population of solutions with other single-solution-based. The same authors noted that these first approaches correspond to proofs-of-concept [17]. Therefore, there many options for investigating approaches to improve or create new and more appropriate approaches.

3   Procedure to Generate Metaheuristics

To find a solution for the metaproblem, a set of primary components must be created. The first component is a container that stores both the feasible current solution for the optimization problem and auxiliary solutions considered during the search process. A second component is a set of functions and terminals to compose the new algorithms to be produced. Such a set contains the typical components of algorithms (functions) as well as tools that allow the construction of a solution for the optimization problem (terminals). From the definition of these elementary components, an initial population of syntax trees can be configured that can evolve into a sequence of later populations using reproduction, crossover, and mutation, which are the proper operators in evolutionary computing [10]. The initial population is generated by the ramped half and half mechanism consisting of randomly generating half the population with full trees up to a default depth and the other half, with partially full trees [28]. The fittest syntax trees are randomly selected and reproduced into the new population. Furthermore, two types of mutations are considered: "point mutation" and "shrink mutation". In the first mutation, a syntax tree node that is a function is replaced by another function that uses the same number of parameters, while in the second, a function node is replaced by any of the functions that act directly on the container where the current solutions are stored. The crossover between two syntax trees is performed by replacing a node of the first syntax tree by a section of the second. The fitness evaluation requires a set of adaptation instances that must be adequately selected. In our case, due to the generated algorithms' stochastic nature, every instance is evaluated m times. The syntax trees finally produced are decoded as algorithms and must also be evaluated externally so that a second set of control instances is required. The set of instances of each type is divided into two groups: the first is used to automatically construct metaheuristics, and the second is used to evaluate the algorithms already built. This evolutionary process is described for t generations in Fig. 1.

The set of functions contains the basic instructions present in any algorithm. They are defined by means of the parameters P_1 and P_2, which are boolean variables, so a "true" indicates that the parameter performed an action; otherwise, it is "false". Specifically, the main functions are $\text{For}(k, P_1, P_2)$, which runs the parameter P_2, whereas P_1 returns "true" and k is the parameter of maximum number of iterations; $\text{And}(P_1, P_2)$ runs the parameter P_1 and P_2 returning "true" if both returned values are also "true"; and $\text{IfThen}(P_1, P_2)$ activates P_2 when running P_1 returns "true".

The set of terminals is divided into two groups. The terminals are based on metaheuristics and terminals that construct a TSP solution. The first terminals contain components of three well-known metaheuristic algorithms: iterated local search (ILS), simulated annealing (SA) and variable neighborhood search (VNS) [36]. These metaheuristics are decomposed in atomic parts, and 12 terminals are obtained, such as terminals based on the computation of the temperature of SA, acceptance criteria of an ILS or SA, and operations in the neighborhood of VNS.

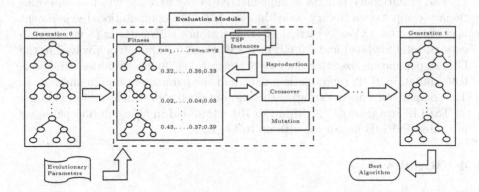

Fig. 1. Process of automatic generation of algorithms.

The following terminals designed for the TSP are based on typical heuristics for the problem and elementary operations to be executed on a solution container:

- ShiftCity: randomly shifts a city of the current solution.
- SwapCities: randomly exchanges two cities of the current solution.
- BlockReverse: randomly reverses the order of the cities of the current solution.
- BlockRule: iteratively reverses randomly the order of the cities of the current solution.
- SwapCitiesRule: iteratively exchanges randomly two cities of the current solution.
- 2-Opt: is an adaptation of a local search procedure for this problem known as 2-opt.

Terminals return true if the action for which they are intended is executed and, false otherwise.

The quality of an algorithm f_a is measured by the relative error in the objective function for a set of instances used during the evolution. In addition, we consider another measure which is the number of obtained solutions that happen to be optimal (also named hits). Let n_l be the number of instances, z_i the value of the optimal solution of the instance i, and u_i the obtained value of the algorithm for instance i. α and β are values that belong to the interval $[0, 1]$ and are used to arbitrarily handle the weight of each term. Furthermore, let hit_i be equal to 1 if the optimal solution is found for the instance i. Then, the evaluation function is represented in Eq. (1). The minimum value of f_a is zero, which means that the algorithm solves all problem instances. We arbitrarily set $\alpha = 0.9$ and $\beta = 0.1$.

$$f_a = \alpha \frac{1}{n_l} \sum_{i=1}^{n_l} \frac{|u_i - z_i|}{z_i} + \beta \frac{n_l - \sum_{i=1}^{n_l} hit_i}{n_l} \qquad (1)$$

The evolutionary process is implemented in the ECJ 25, which is an evolutionary computation library coded in Java [20]. The computational experiment is performed on a Google Virtual Machine instance with 2.0 GHz (8 virtual processors, Intel Skylake) and 7.0 GB of RAM. Genetic parameters involved in the GP are population size, 100; number of generations, 30; probabilities: crossover, 0.85; mutation: 0.10; reproduction: 0.05, and the parameters $k = 15$ and $m = 5$. The comparison (called testing) of the best-obtained algorithm is performed with 14 TSPLIB instances [30], with up to 101 cities, and in the evolution phase we use three TSPLIB instances, with up to 58 cities.

4   Results

The best algorithm found finds near-optimal solutions for the 14 evaluation instances. The algorithm was obtained after performing three runs with the same parameter values, changing only the seed. Consequently, 3,000 combinations were inspected. In turn, to evaluate the algorithm, ten runs were performed with each instance, and the resulting values are presented in Table 1. The name of the instance is described in the first column, the second column contains the minimum relative error value with respect to the optimal solution, while in the third column, the average relative error of all runs is presented. In the last column, the average computational time of the algorithm with each instance is reported.

The result suggests that it is feasible to combine different elementary components of the various metaheuristics and assemble such components appropriately to face the TSP. As t is observed in Table 1 in the ten instances, the optimal solution was obtained in at least one of the ten runs for each instance. In the ten runs, the optimal solution of the brazil58 instance was obtained. Note that the average computational time was between 0.47 and 3.3 s.

The structure of one of the best generated algorithms corresponds to a variant of the ILS algorithm. Algorithm 1 has four stages. In the first stage (lines 10–15), the algorithm performs a local search, and in the second stage, it verifies the acceptance or rejection of the solution found by a local search process (lines 16–21). In the third stage, a perturbation is performed (line 22), and in the fourth stage, the algorithm ends with a new local search (line 24–26). The algorithm considers two stop criteria explicitly established in lines 27 and 29. Line 27 establishes that if the first local search does not improve the current solution, the algorithm stops. Line 29 stops the algorithm according to the number of iterations predefined as the instance size. The generated metaheuristic differs from ILS in the order of the instructions. The algorithms also incorporate elements of SA that do not contribute to the TSP solution; these are instructions that work as bloating code that is common when using GP.

Table 1. Performance of generated metaheuristic on 14 instances.

Instances	Min (%)	Avg (%)	Avg time (Sec)
eil51	0.23	1.03	0.47
berlin52	0.00	3.48	0.50
brazil58	0.00	0.00	0.69
st70	0.00	1.10	1.21
eil76	0.00	1.21	1.53
pr76	0.00	0.37	1.48
rat99	0.00	1.73	3.44
kroA100	0.00	0.43	3.42
kroB100	0.00	1.01	3.78
kroC100	0.00	0.44	3.30
kroD100	0.07	1.24	3.34
kroE100	0.00	0.50	3.45
rd100	0.85	1.75	3.32
eil101	0.32	1.21	3.55
Average	0.11	1.11	2.39

Algorithm 1. Generated algorithm

```
1: function FUNCTION
2:     if 2-opt() then
3:         for 1 to k do
4:             2-opt();
5:             2-opt();
6:         end for
7:     end if
8: end function
9: repeat
10:     γ ← false;
11:     δ ← false;
12:     for 1 to k and FUNCTION() = true do
13:         γ ← true;
14:         LinearCooling();
15:     end for
16:     if γ = true and 2-opt() then
17:         for 1 to k and LogCooling() do
18:             δ ← true;
19:             Deterministic_ChooseIfBetter();
20:         end for
21:     end if
22:     if γ = true and δ = true and SwapCities() then
23:         if 2-opt() then
24:             FUNCTION();
25:         end if
26:     else
27:         break;
28:     end if
29: until iter = n
```

The extension of AGA to automatically produce metaheuristics produced similar results to those found when AGA was used to produce specific heuristics for various optimization problems. New metaheuristics were produced, as combination of the initially defined components of well-known metaheuristics. In addition, such metaheuristics produce near-optimal solutions for at least a small group of instances. It is clear that the scalability of this result requires more extensive experimentation. Likewise, the generated algorithms must be properly parameterized to obtain a better computational performance, a process that could also be included automatically. An interesting consequence of the result is that the automation process adopted significantly accelerates research in this field. Many algorithmic combinations can be explored with low computational effort; in fact, the reported experiment lasted 158 min.

5   Conclusions

This work described a process for the automatic design of metaheuristics for TSP. The algorithms were produced by automatic generation of algorithms through genetic programming from a set of elementary components. In particular, we used terminals based on the metaheuristic algorithms ILS, SA, and VNS. The resulting algorithms are combinations of existing metaheuristics, and the best algorithm found is a variant of ILS. However, the generated metaheuristic provides good performance in terms of the solution quality in a very short computational time. Future research will focus on improving the performance of the generated metaheuristics and combining the different components of the metaheuristics in a better way. Additional future research will focus on extending the proposed method to other variants of the TSP or even other optimization problems.

Acknowledgment. This research was partially funded by the Complex Engineering Systems Institute (CONICYT PIA/BASAL AFB180003). We also acknowledge projects USA1899–Vridei 061919VP-PAP Universidad de Santiago de Chile, DICYT-USACH 061919PD, and VRID INICIACIÓN 220.097.016-INI, Vicerrectoría de Investigación y Desarrollo (VRID), Universidad de Concepción.

References

1. Acevedo, N., Rey, C., Contreras-Bolton, C., Parada, V.: Automatic design of specialized algorithms for the binary knapsack problem. Expert Syst. Appl. **141**, 112908 (2020)
2. Alfaro-Fernández, P., Ruiz, R., Pagnozzi, F., Stützle, T.: Automatic algorithm design for hybrid flowshop scheduling problems. Eur. J. Oper. Res. **282**(3), 835–845 (2020)
3. Bertolini, V., Rey, C., Sepúlveda, M., Parada, V.: Novel methods generated by genetic programming for the guillotine-cutting problem. Sci. Program. **2018**, 1–13 (2018)

4. Blum, C., Puchinger, J., Raidl, G.R., Roli, A.: Hybrid metaheuristics in combinatorial optimization: a survey. Appl. Soft Comput. **11**(6), 4135–4151 (2011)
5. Blum, C., Raidl, G.R.: Hybrid Metaheuristics: Powerful Tools for Optimization. Artificial Intelligence: Foundations, Theory, and Algorithms, Springer, Cham (2016). https://doi.org/10.1007/978-3-319-30883-8
6. Conforti, M., Cornuéjols, G., Zambelli, G.: Integer Programming. Graduate Texts in Mathematics, vol. 271. Springer, Cham (2014). https://doi.org/10.1007/978-3-319-11008-0
7. Contreras-Bolton, C., Gatica, G., Parada, V.: Automatically generated algorithms for the vertex coloring problem. PLoS ONE **8**(3), e58551 (2013)
8. Contreras-Bolton, C., Parada, V.: Automatic design of algorithms for optimization problems. In: 2015 Latin America Congress on Computational Intelligence (LACCI), Curitiba, Brazil, pp. 1–5. IEEE (2015)
9. Contreras-Bolton, C., Rey, C., Ramos-Cossio, S., Rodríguez, C., Gatica, F., Parada, V.: Automatically produced algorithms for the generalized minimum spanning tree problem. Sci. Program. **2016**, 11 (2016)
10. Eiben, A., Smith, J.: Introduction to Evolutionary Computing. Natural Computing Series, 2nd edn. Springer, Heidelberg (2015). https://doi.org/10.1007/978-3-662-44874-8
11. Gendreau, M., Potvin, J.Y. (eds.): Handbook of Metaheuristics. International Series in Operations Research & Management Science, vol. 272. Springer, Cham (2019). https://doi.org/10.1007/978-3-319-91086-4
12. Hassan, A., Pillay, N.: A meta-genetic algorithm for hybridizing metaheuristics. In: Oliveira, E., Gama, J., Vale, Z., Lopes Cardoso, H. (eds.) EPIA 2017. LNCS (LNAI), vol. 10423, pp. 369–381. Springer, Cham (2017). https://doi.org/10.1007/978-3-319-65340-2_31
13. Hassan, A., Pillay, N.: Hybrid metaheuristics: an automated approach. Expert Syst. Appl. **130**, 132–144 (2019)
14. Kolker, A.: Healthcare Management Engineering: What Does This Fancy Term Really Mean? Springer, New York (2012). https://doi.org/10.1007/978-1-4614-2068-2
15. Korte, B., Vygen, J.: Combinatorial Optimization: Theory and Algorithms. Algorithms and Combinatorics, vol. 21, 6th edn. Springer, Heidelberg (2018). https://doi.org/10.1007/978-3-662-56039-6
16. López-Ibáñez, M., Dubois-Lacoste, J., Pérez Cáceres, L., Birattari, M., Stützle, T.: The irace package: iterated racing for automatic algorithm configuration. Oper. Res. Perspect. **3**, 43–58 (2016)
17. López-Ibáñez, M., Kessaci, M.E., Stützle, T.: Automatic design of hybrid metaheuristics from algorithmic components. Technical report, TR/IRIDIA/2017-012, IRIDIA, Université Libre de Bruxelles, Belgium (2017)
18. López-Ibáñez, M., Stutzle, T.: The automatic design of multiobjective ant colony optimization algorithms. IEEE Trans. Evol. Comput. **16**(6), 861–875 (2012)
19. Loyola, C., Sepúlveda, M., Solar, M., Lopez, P., Parada, V.: Automatic design of algorithms for the traveling salesman problem. Cogent Eng. **3**(1), 1255165 (2016)
20. Luke, S.: ECJ then and now. In: Proceedings of the Genetic and Evolutionary Computation Conference Companion on - GECCO 2017, pp. 1223–1230. ACM Press, New York (2017)
21. Marmion, M.-E., Mascia, F., López-Ibáñez, M., Stützle, T.: Automatic design of hybrid stochastic local search algorithms. In: Blesa, M.J., Blum, C., Festa, P., Roli, A., Sampels, M. (eds.) HM 2013. LNCS, vol. 7919, pp. 144–158. Springer, Heidelberg (2013). https://doi.org/10.1007/978-3-642-38516-2_12

22. Pagnozzi, F., Stützle, T.: Automatic design of hybrid stochastic local search algorithms for permutation flowshop problems. Eur. J. Oper. Res. **276**(2), 409–421 (2019)
23. Pagnozzi, F., Stützle, T.: Evaluating the impact of grammar complexity in automatic algorithm design. Int. Trans. Oper. Res. (2020, for forthcoming)
24. Pagnozzi, F., Stützle, T.: Automatic design of hybrid stochastic local search algorithms for permutation flowshop problems with additional constraints. Oper. Res. Perspect. **8**, 100180 (2021)
25. Papadimitriou, C.H., Steiglitz, K.: Combinatorial Optimization: Algorithms and Complexity. Prentice-Hall Inc., Upper Saddle River (1982)
26. Parada, L., Herrera, C., Sepúlveda, M., Parada, V.: Evolution of new algorithms for the binary knapsack problem. Nat. Comput. **15**(1), 181–193 (2015). https://doi.org/10.1007/s11047-015-9483-8
27. Pétrowski, A., Ben-Hamida, S.: Evolutionary Algorithms. Wiley, Hoboken (2017)
28. Poli, R., Langdon, W.B., Mcphee, N.F.: A Field Guide to Genetic Programming. Lulu Enterprises, UK Ltd (2008)
29. Raidl, G.R., Puchinger, J., Blum, C.: Metaheuristic hybrids. In: Gendreau, M., Potvin, J.-Y. (eds.) Handbook of Metaheuristics. ISORMS, vol. 272, pp. 385–417. Springer, Cham (2019). https://doi.org/10.1007/978-3-319-91086-4_12
30. Reinelt, G.: TSPLIB-a traveling salesman problem library. ORSA J. Comput. **3**(4), 376–384 (1991)
31. Ryser-Welch, P., Miller, J.F., Asta, S.: Generating human-readable algorithms for the travelling salesman problem using hyper-heuristics. In: Proceedings of the Companion Publication of the 2015 on Genetic and Evolutionary Computation Conference - GECCO Companion 2015, pp. 1067–1074. ACM Press, New York (2015)
32. Schrijver, A.: Theory of Linear and Integer Programming. Wiley, Hoboken (1998)
33. Silva-Muñoz, M., Contreras-Bolton, C., Semaan, G.S., Villanueva, M., Parada, V.: Novel algorithms automatically generated for optimization problems. In: 2019 38th International Conference of the Chilean Computer Science Society (SCCC), pp. 1–7 (2019)
34. Silver, E.A., Pyke, D.F., Thomas, D.J.: Inventory and Production Management in Supply Chains, 4th edn. Taylor & Francis, Boca Raton (2016)
35. Stützle, T., López-Ibáñez, M.: Automated design of metaheuristic algorithms. In: Gendreau, M., Potvin, J.-Y. (eds.) Handbook of Metaheuristics. ISORMS, vol. 272, pp. 541–579. Springer, Cham (2019). https://doi.org/10.1007/978-3-319-91086-4_17
36. Talbi, E.G.: Metaheuristics: From Design to Implementation. Wiley, Oxford (2009)
37. Tezel, B.T., Mert, A.: A cooperative system for metaheuristic algorithms. Expert Syst. Appl. **165**, 113976 (2021)
38. Ting, T.O., Yang, X.-S., Cheng, S., Huang, K.: Hybrid metaheuristic algorithms: past, present, and future. In: Yang, X.-S. (ed.) Recent Advances in Swarm Intelligence and Evolutionary Computation. SCI, vol. 585, pp. 71–83. Springer, Cham (2015). https://doi.org/10.1007/978-3-319-13826-8_4
39. Toth, P., Vigo, D.: Vehicle Routing: Problems, Methods, and Applications, 2nd edn. Society for Industrial and Applied Mathematics, Philadelphia (2014)
40. Wang, J.: Management Science, Logistics, and Operations Research. Advances in Logistics, Operations, and Management Science. IGI Global (2014)
41. Wright, M.: Operational Research Applied to Sports. Palgrave Macmillan UK, London (2015)

Discrete PSO Strategies for Search in Unit Lattice of m-dimensional Simplex

Marek Kvet[✉] and Jaroslav Janáček

Faculty of Management Science and Informatics, University of Žilina, Univerzitná, 8215/1,
010 26 Žilina, Slovakia
{marek.kvet,jaroslav.janacek}@fri.uniza.sk

Abstract. The path-relinking based strategies have proved to be very powerful tool for designing the emergency service systems by deploying a given number of service centers in a finite set of possible center locations. Nevertheless, if the original approach to the emergency service system design is generalized to the case, when more than one facility can be placed at the same possible center location, the question emerges whether the generalized version of the path-relinking method is able to keep its former efficiency. It must be taken into account that the generalized path-relinking method performs its search in nodes of integer lattice of an m-dimensional simplex instead of in a sub-set of unit hypercube vertices. This generalization may considerably change characteristics of the path-relinking based searching strategies. This contribution is devoted to studying and comparing two original particle swarm strategies called the shrinking fence and spider search strategies, which employ the generalized path-relinking method.

Keywords: Emergency medical service system · Heuristics · Generalized path-relinking method · Discrete PSO strategies

1 Introduction

Applied Informatics belongs to one of the currently fastest developing scientific fields. It deals with creation, collection, processing, storage, transformation, access and usage of any kind of information in natural and artificial, general and special systems. Its content aims at the properties and methods of information processing in terms of their optimal availability and usability. It has purely scientific (theoretical) components that examine the subject regardless of the application, and application (practical) components that contribute to the development of services and products. In this paper, we focus on applying the knowledge of Applied Informatics and programming into the specific family of Operations Research problems. We pay attention to the problem of designing and optimizing a network of rescue service stations in a middle sized geographical region [1, 17, 19]. In other words presented research deals with certain type of location problems, for which suitable heuristics are being developed and studied.

Generally, the locations problems may be divided into two independent groups – they can be either continuous or discrete. When talking about finding the optimal service

© Springer Nature Switzerland AG 2022
B. Dorronsoro et al. (Eds.): META 2021, CCIS 1541, pp. 59–69, 2022.
https://doi.org/10.1007/978-3-030-94216-8_5

center deployment for Emergency Medical Service (EMS), then we have to consider such a fact, that the service centers cannot be located anywhere due to certain requirements given by law. Therefore, the problem of finding the optimal locations of EMS stations is usually formulated as median-type problem, which has been recently studied by many researchers [2, 4, 5, 7, 13, 18].

The simplest median-based model is the weighted p-median problem with a wide spectrum of applications. Since it belongs to well-known and commonly used optimization models, several authors have analyzed the possibilities of its fast solving either by exact or approximate and heuristic methods [1, 6, 14].

Under the assumption that the service is not possible to be provided to more than one patient simultaneously by the same staff, the EMS systems operates as a queuing system. Obviously, when life is directly endangered or health gets suddenly worse, the rescue service is provided by such a facility, which is the nearest available one. From this point of view, the concept of so-called generalized disutility can be used [10, 12, 15].

Furthermore, if we look at an existing EMS system, we can observe that there are more than one facilities and staff located at some service center locations. Thus, we should consider this feature when a mathematical model of the problem is being formulated [8]. Of course, such an original model modification makes the problem more complex and possible usage of common exact and heuristic approaches designed for the median-type problems is questionable.

Within this paper, we study the path-relinking based strategies, which have proved to be very powerful tool for designing the emergency service systems by deploying a given number of service centers in a finite set of possible center locations. The main goal of presented research consists in answering the question whether the generalized version of the path-relinking method is able to keep its former efficiency. It must be taken into account that the generalized path-relinking method performs its search in nodes of integer lattice of an m-dimensional simplex instead of exploring a sub-set of unit hypercube vertices. This generalization may considerably change characteristics of the path-relinking based searching strategies. Therefore, we concentrate on two original Particle Swarm Optimization (PSO) strategies called the shrinking fence and spider search strategies, which employ the generalized path-relinking method [11, 16]. To study suggested heuristic approaches, a computational study with real world middle-sized problem instances was performed and the obtained results are reported in a separate section.

2 Generalized p-facility Location Problem and Path-Relinking Search

The generalized p-facility location problem is formulated as a task to deploy p facilities in m network nodes so that the mean distance between a user and the nearest available facility is minimal. It is assumed that the system of facilities services demands of n users located also at nodes of the transportation network. A user j generates his demand randomly with frequency b_j. As the system processes the demands for service similarly to a queuing system equipped with p service lines, a current demand is assigned to the nearest available facility, which need not be the closest one. For each user location, r nearest facilities is taken into account for the demand satisfaction and a sequence $q_1, ..., q_r$ probability values is considered, where q_k is probability that the k-th nearest facility to the user is the first available one. Unlike the previous approaches, here we admit that more than one facility can be located in one node of the network. Assuming that d_{ij} denotes the network distance between network nodes i and j the generalized p-facility location problem can be described by the following integer programming model, in which a series of integer location variables $x_i \in Z^+$ will be introduced for each $i = 1, ..., m$, to model the number of facilities located at the node i.. The value of variable y_i gives the number of facilities located at location i. In addition, a series of allocation variables $w_{ijk} \in \{0, 1\}$ will be introduced for $i = 1, ..., m, j = 1, ..., n$ and $k = 1, ..., r$, where $w_{ijk} = 1$ if user demand emerged at j is assigned to a service node i for the k-th nearest facility.

$$Minimize \quad \sum_{j=1}^{n} b_j \sum_{k=1}^{r} q_k \sum_{i=1}^{m} d_{ij} w_{ijk} \tag{1}$$

$$Subject\ to \quad \sum_{i=1}^{m} x_i = p \tag{2}$$

$$\sum_{i=1}^{m} w_{ijk} = 1 \ for\ j = 1, ..., m,\ k = 1, ..., r \tag{3}$$

$$\sum_{k=1}^{r} w_{ijk} \leq x_i \ for\ j = 1, ..., n,\ i = 1, ..., m \tag{4}$$

$$x_i \in Z^+ \ for\ i = 1, ..., m \tag{5}$$

$$w_{ijk} \in \{0, 1\} \ for\ i = 1, ..., m,\ j = 1, ..., n,\ k = 1, ..., r \tag{6}$$

The formula (1) expresses the sum of mean distances from users' locations to the nearest available facility location. As the sequence of $\{q_k\}$ is decreasing, a demand of a user's location j is assigned to the nearest facility location i for $k = 1$. Similarly, the demand will be assigned to the second nearest facility for the case $k = 2$, etc.

Constraint (2) determines the number of deployed facilities. Series of constraints (3) ensures that demand at location j can be allocated to exactly one facility for the case k. This means that the k-th nearest facility is the first available one. Series of constraints (4) enables to assign a demand at user's location j to a possible service node i at most x_i times.

The problem (1)–(6) is more complex than the case, when only one facility can be located at a service node. The study reported in [8] has showed that computational time necessary to solve the problem (1)–(6) to optimality using an IP-solver exceeded an acceptable limit. This finding approves usage of a heuristic approaches to the problem solution. We were inspired by discrete particle swarm optimization algorithms [3, 20], which proved to be an efficient tool for this kind of p-location problem, but without the possibility to place more than one facility at the same service node.

The mentioned algorithms [11, 16] called the shrinking fence and the spider search are based on systematic examination of a series of the shortest paths connecting pairs of feasible hypercube vertices in a surface of the unit hypercube. To be able to use the above mentioned searching strategies for heuristic solution of the problem (1)–(6), we suggested a new version of the path-relinking method, which is able to examine the shortest path between two nodes of a unit lattice of an m-dimensional simplex.

A feasible solution of (1)–(6) is described by an m-dimensional vector $\mathbf{x}$ with integer non-negative components, sum of which equals to p. The shortest path between two feasible solutions has a length, which equals to Manhattan distance of the two vectors. The suggested path-relinking method proceeds according to the following steps.

FacetPathRelinking($\mathbf{x}$, $\mathbf{y}$)

0. Initialize $\mathbf{x}^{res} = argmin\{f(\mathbf{x}), f(\mathbf{y})\}$. Define sets M^+ and M^- of component indices by prescriptions $M^+ = \{i = 1, \ldots, m : x_i < y_i\}$ and $M^- = \{i = 1, \ldots, m : x_i > y_i\}$.
1. If $\rho(\mathbf{x}, \mathbf{y}) > 2$ perform step 2, otherwise return $\mathbf{x}^{res}$ and terminate.
2. Find $[u, v] \in M^+ \times M^-$ using the definition
    $$[u, v] = argmin\{f(\text{exchange}(\mathbf{x}, i, j)) : [i, j] \in M^+ \times M^-\}$$ and perform operations
    $\mathbf{x} = \text{exchange}(\mathbf{x}, u, v))$; if $x_u = y_u$; if $x_u = y_u$, then $M^+ = M^+ - \{u\}$; if $x_v = y_v$ then $M^- = M^- - \{v\}$ $\mathbf{x}^{res} = argmin\{f(\mathbf{x}^{res}), f(\mathbf{x})\}$.
    Having performed the above adjustments, exchange $\mathbf{x}$ with $\mathbf{y}$ and M^+ with M^- and go to step 1.

Comments: In the above algorithm, $\rho(\mathbf{x}, \mathbf{y})$ denotes the Manhattan distance of $\mathbf{x}$ and $\mathbf{y}$ defined by (7).

$$\rho(\mathbf{x}, \mathbf{y}) = \sum_{i=1}^{m} |x_i - y_i| \qquad (7)$$

The operation exchange($\mathbf{x}, u, v$) for $u \in M^+$ and $v \in M^-$ issues the vector $\underline{\mathbf{x}}$, components of which are defined by the following substitutions $\underline{x}_i = x_i$ for $i = 1, \ldots, m, i \neq u$, $i \neq v$ and $\underline{x}_u = x_u + 1$, $\underline{x}_v = x_v - 1$.

The value of function $f(\mathbf{x})$ for a given $\mathbf{x}$ is computed according to (1)–(6) after fixing the values of x_i for $i = 1, \ldots, m$.

The algorithm *FacetPathRelinking*($\mathbf{x}$, $\mathbf{y}$) examines the shortest path connecting integer points $\mathbf{x}$ and $\mathbf{y}$ in an m-1 dimensional facet of simplex determined by (2) and (5). The value $\rho(\mathbf{x}, \mathbf{y})$ is obviously even integer and every performance of the step 2 reduces this distance by two. Thanks to the exchange $\mathbf{x}$ and $\mathbf{y}$ at the end of step 2, the algorithm constructs and examines the path alternately from the both ends.

3 Particle Swarm Strategies Based on Path-Relinking Method

Principles of the further applied strategies were obtained from [11, 16] and adapted for the search in the set of feasible solutions of the problem (1)–(6) using the above suggested version of the path-relinking method. The both proposed algorithms start with an initial swarm S of input solutions-particles and use the path-relinking method as a function *FacetParthRelinking*($\mathbf{x}$, $\mathbf{y}$), which returns the best-found-solution in the examined path.

The shrinking fence algorithm imitates building and maintaining a fence, which surrounds a herd of solutions. At the beginning, posts of the fence are represented by known solutions of the initial set S. The order of posts in the fence is given by ordering of the associated solutions according their objective function values. It is assumed that the neighboring posts are connected by fence parts. During the optimization process, the individual fence parts are examined, a new, better position of a post is found and the new post replaces one of the neighboring posts. If one of the neighboring posts is closer to the new one more than a given distance, then the unnecessary post is removed. The best-found solution obtained by the inspections of fence parts is output of the algorithm.

The shrinking fence algorithm follows.

0. {Building up phase}

   Order the solutions of input swarm S increasingly by their objective function values. This way, create a sequence s^0, ..., $s^{|S|-1}$. Initialize the best-found solution $\mathbf{x}^{best} = s^0$ and the set of new posts $\underline{S}$ by empty set $\emptyset$.

1. {Maintenance phase}

   For $t = |S| - 1, \ldots, 1$, inspect the fence part connecting the posts s^t and s^{t-1} and define the new post position $\mathbf{x}^{new}$ by $\mathbf{x}^{new} = $ *FacetPathRelinking*(s^t,s^{t-1}). If $\rho(s^t, \mathbf{x}^{new}) < d^{min}$, then put $\underline{S} = \underline{S} \cup \{\mathbf{x}^{new}\}$. Replace the best-found solution by $\mathbf{x}^{best} = argmin\{f(\mathbf{x}^{best}), f(\mathbf{x}^{new})\}$.

   Inspect the fence part connecting s^0 and $s^{||s|-1}$ by $\mathbf{x}^{new} = $ *FacetPathRelinking*($s^0,s^{||s|-1}$)

   If $\rho(s^0, \mathbf{x}^{new}) < d^{min}$, then put $\underline{S} = \underline{S} \cup \{\mathbf{x}^{new}\}$. Replace the best-found solution by $\mathbf{x}^{best} = argmin\{f(\mathbf{x}^{best}), f(\mathbf{x}^{new})\}$.

   {Improving process controlling}

   If the termination condition is fulfilled, then terminate and return $\mathbf{x}^{best}$. Otherwise, update $S = \underline{S}$, reorder the elements of S according to increasing objective function values and put $\underline{S} = \emptyset$. Go to step 1.

   Comment: The termination condition consists of two clauses. The process is terminated whenever the number of updates of the set S reaches the limit maxPop or if the expended computational time exceeds the threshold maxTime.

The next presented heuristic called the spider search is evoked by spider's web creation, which starts with linking fixed points with a center of the web by spider's thread and subsequent linking of the neighboring fixed points. Then some inner web nodes are established and the linking process to web center and then the mutual connections of the neighboring web nodes continue up to the moment, when the web is dense enough.

0. Initialize the starting swarm by a set S and order the solutions increasingly according to their objective function values into the sequence $s^0, \ldots, s^{|S|-1}$ Initialize the best-found solution $\mathbf{x}^{center} = s^0$.

1. Process the swarm $\{s^0, \ldots, s^{|S|-1}\}$ and web center $\mathbf{x}^{center}$ in the following way: Update the web center by $\mathbf{x}^{center} = argmin\{f(FacetPathRelinking(\mathbf{x}^{center}, s^t))$:

$t = 1, \ldots, |S| - 1$} and insert the final $\mathbf{x}^{center}$ into the new swarm. For $t = 1, \ldots, |S| - 1$, determine $\mathbf{x}^* = FacetPathRelinking\left(\mathbf{s}^t, \mathbf{s}^{t-1}\right)$ and if there is no identical solution, insert $\mathbf{x}^*$ into the new swarm, otherwise skip the insertion. Finally perform $\mathbf{x}^* = FacetPathRelinking\left(\mathbf{s}^0, \mathbf{s}^{|S|-1}\right)$ and add $\mathbf{x}^*$ to the new swarm.

2. If the termination condition is fulfilled, then the solving process finishes with the output defined by the best-found solution. Otherwise reorder new swarm, determine new web center $\mathbf{x}^{center}$ and go to step 1.

Comment: The termination condition consists of two clauses. The process is terminated whenever the number of the swarm updates reaches the limit maxPop or if the expended computational time exceeds the threshold maxTime.

4   Computational Experiments

The main goal of performed computational study was to verify the efficiency of suggested discrete PSO strategies for search in unit lattice of m-dimensional simplex. Note that mentioned heuristics were originally developed and designed for a simple version of the weighted p-median problem [11, 16]. Therefore, their quality characteristics may change when the original model gets a more general form (1)–(6).

The numerical experiments reported in this paper were performed on a notebook equipped with the Intel® Core™ i7 3610QM 2.3 GHz processor and 8 GB of memory. The presented algorithms were implemented in the Java language making use of the NetBeans IDE 8.2 environment.

As far as the problem instances used in this computational study are concerned, they originate from real EMS system, which is operated in eight regions of Slovakia. The problem instances were used also in our previous research activities, the results of which are available in [9–11, 16] and in many others. The cardinalities of the set of possible service center locations and the set of system users vary from 87 to 664 locations. The organization of the Slovak self-governing regions is depicted in Fig. 1.

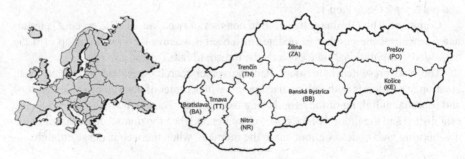

Fig. 1. Used benchmarks – self-governing regions of Slovakia.

The parameters of individual benchmarks are summarized in the following Table 1. in which also the exact solutions taken from another research [8] are reported. The coefficients $q_k, k = 1, \ldots, r$ for $r = 3$ stand for probabilities that the k-th nearest service center is the closest available one. The values of these coefficients were set so that $q_1 = 0{,}77063$, $q_2 = 0{,}16476$ and $q_3 = 1\text{-}q_1\text{-}q_2$. These values were obtained from a simulation model of existing EMS system in Slovakia published in [12].

The first four columns of Table 1 contain the basic characteristics of used problem instances. Column denotations keep the same meaning as used in the model (1)–(6). The last column of the table denoted by *OptObjF* is used to report the objective function value of the exact optimal solution of the model (1)–(6), which was computed in previous research reported in [8].

Table 1. Basic benchmarks characteristics and the optimal objective function values

Region	m	N	p	OptObjF
BA	87	87	25	18450
BB	515	515	46	38008
KE	460	460	38	40711
NR	350	350	36	40987
PO	664	664	44	46884
TN	276	276	26	31260
TT	249	249	22	36401
ZA	315	315	36	36929

An individual experiment was organized so that both compared PSO strategies, i.e. the shrinking fence and the spider search employing the generalized path-relinking method were applied to obtain the result of the problem described by the mathematical model (1)–(6). Since the optimal objective function value is available, the suggested algorithms can be compared from the viewpoint of solution accuracy.

Before reporting the achieved results, it must be noted that the basic idea of both solving approaches follows from the fact that the individual strategy starts from a set of feasible solutions, which can be provided by so-called uniformly deployed set. This set can be constructed independently on the solved instance. The process of a uniformly deployed set construction is reported in [9] and its possible usage can be found for example in [10, 11, 16]. The common property of a uniformly deployed set is that an arbitrary permutation of the locations generates a new uniformly deployed set with the same characteristics. We used this property to obtain ten different starting sets for each self-governing region presented in Table 1 and the values plotted in further Table 2 were obtained by averaging ten problem instances. The original uniformly deployed sets of zero-one solutions obtained from [9, 10] were adjusted by a greedy process to include some initial solutions outside the unit m-dimensional hypercube. Both suggested methods were run for stopping rule parameters maxPop = 8 and maxTime = 120 s.

The following Table 2 contains the average results. The structure of the table is formed by two parts – separate for each studied PSO strategy. For each heuristic approach we report the objective function value *ObjF* and the computational time *CT* in seconds.

Table 2. Comparison of discrete PSO strategies for the generalized weighted p-median problem – average results of ten runs with different uniformly deployed sets of solutions

Region	Shrinking fence		Spider search	
	ObjF	*CT*	*ObjF*	*CT*
BA	18751	1.12	18730	2.84
BB	39924	147.68	38094	211.21
KE	40711	92.46	40715	139.55
NR	41062	30.57	41062	58.59
PO	56416	124.73	47005	132.21
TN	31568	16.19	31540	37.28
TT	36768	10.44	36750	21.77
ZA	37030	27.81	37028	51.95

For completeness of reported results, we provide the readers with one additional Table 3, which contains the detailed results for the self-governing region of Žilina. Table 3 has the same structure as the former Table 2.

Table 3. Comparison of discrete PSO strategies for the generalized weighted p-median problem –results of ten runs with different uniformly deployed sets of solutions for the self-governing region of Žilina

Run	Shrinking fence		Spider search	
	ObjF	*CT*	*ObjF*	*CT*
1	36929	28.32	36929	51.68
2	36929	28.28	36929	53.34
3	36929	27.91	36929	51.86
4	37848	27.58	37828	52.78
5	36929	27.74	36929	51.25
6	36929	28.00	36929	53.00
7	36929	27.69	36929	50.36
8	36993	27.44	36993	52.35
9	36929	27.63	36929	51.46
10	36964	27.46	36964	51.37

All reported results indicate that the quality of obtained results is very satisfactory. From the point of solution accuracy, the strategy of a spider search seems better, because the average gap from the optimal objective function value achieves only the value of 0.54% while the first studied shrinking fence strategy brings worse results. As far as the computational time is concerned, both strategies can achieve the result in acceptably short time and can be used to solve practical real world problems.

5 Conclusions

This contribution was focused on two strategies employing the path-relinking method. The main research goal was aimed at the finding, whether the adjusted shrinking fence and spider search strategies are able to prove the same efficiency as their simple original versions when used for the p-location problem solution subject to the assumption that more than one facility can be located at the same possible service center location.

Suggested methods are based on the path-relinking method and they make use of previously developed search strategies. The novelty of presented original method extension consists in adjusting the heuristics for different space, in which the solutions are being explored. It must be realized that the mathematical problem formulation, to which the suggested heuristics were adjusted, makes use of the concept of generalized disutility, which assumes, that the service does not have to be provided by the nearest located service center, because it may be temporarily unavailable. In such a case, the request for rescue service is assigned to the nearest available center. The second modification of the original model consists in significant variables definition scope extension. It means that more than one facilities are allowed to be located in the same possible service center locations. This way, the former binary decision variables change into integers, what can make many available solving tool necessary to be adjusted or rebuilt.

The reported results of numerical experiments aimed at heuristic solving techniques for the multiple p-facility location problems with the generalized objective function show that the suggested strategies keep their useful features and both of them can be used for effective solving middle-sized problem instances. The accuracy of the resulting solution is satisfactory and the resulting system design can be obtained in acceptably short computational time. Based on performed numerical experiments we can conclude that we have constructed a very fast and effective heuristic approach to the generalized p-location problems.

Future research in this scientific field could be concentrated on rules, which would enable to reduce the starting set of p-location problem solutions and on developing other search strategies, which could improve the studied characteristic of the heuristic solving approach.

Acknowledgment. This work was supported by the research grants VEGA 1/0089/19 "Data analysis methods and decisions support tools for service systems supporting electric vehicles", VEGA 1/0689/19 "Optimal design and economically efficient charging infrastructure deployment for electric buses in public transportation of smart cities", and VEGA 1/0216/21 "Design of emergency systems with conflicting criteria using artificial intelligence tools". This work was supported by the Slovak Research and Development Agency under the Contract no. APVV-19-0441.

References

1. Avella, P., Sassano, A., Vasil'ev, I.: Computational study of large scale p-median problems. Math. Program. **109**, 89–114 (2007)
2. Current, J., Daskin, M., Schilling, D.: Discrete network location models, Drezner, Z., et al. (ed.) Facility Location: Applications and theory, Springer, pp. 81–118 (2002). https://doi.org/10.1007/978-3-642-56082-8_3
3. Davendra, D., Zelinka, I.: Self-Organizing Migrating Algorithm, Methodology and Implementation. Springer, Studies in Computational Intelligence, p. 289 (2016). https://doi.org/10.1007/978-3-319-28161-2
4. Doerner, K.F., Gutjahr, W.J., Hartl, R.F., Karall, M., Reimann, M.: Heuristic solution of an extended double-coverage ambulance location problem for Austria. CEJOR **13**(4), 325–340 (2005)
5. Drezner, T., Drezner, Z.: The gravity p-median model. Eur. J. Oper. Res. **179**, 1239–1251 (2007)
6. García, S., Labbé, M., Marín, A.: Solving large p-median problems with a radius formulation. INFORMS J. Comput. **23**(4), 546–556 (2011)
7. Gendreau, M., Potvin, J.-Y. (eds.): Handbook of metaheuristics. ISORMS, vol. 272. Springer, Cham (2019). https://doi.org/10.1007/978-3-319-91086-4
8. Janáček, J.: Multiple p-facility location problem with randomly emerging demands. In: Strategic Management and its Support by Information Systems 2021: 14th International Conference, Technical University of Ostrava, in print (2021)
9. Janáček, J., Kvet, M.: Uniform deployment of the p-location problem solutions. In: Operations Research Proceedings 2019: Selected Papers of the Annual International Conference of the German Operations Research Society (GOR), Dresden, Germany, September 4–6, 2019: Springer, 2020 (2019), ISBN 978-3-030-48438-5, ISSN 0721-5924, pp. 315–321
10. Janáček, J., Kvet, M.: Usage of uniformly deployed set for p-location min-sum problem with generalized disutility. In: SOR 2019 proceedings, pp. 494–499 (2019)
11. Janáček, J., Kvet, M.: Shrinking fence search strategy for p-location problems. In: CINTI 2020: IEEE 20th International Symposium on Computational Intelligence and Informatics, Budapest, pp. 55–60 (2020)
12. Jankovič, P.: Calculating reduction coefficients for optimization of emergency service system using microscopic simulation model. In: 17th International Symposium on Computational Intelligence and Informatics, pp. 163–167 (2016)
13. Jánošíková, Ľ., Žarnay, M.: Location of emergency stations as the capacitated p-median problem. In: Quantitative Methods in Economics (Multiple Criteria Decision Making XVII), pp. 117–123 (2014)
14. Kozel, P. Orlíková, L., Pomp, M., Michalcová, Š.: Application of the p-median approach for a basic decomposition of a set of vertices to service vehicles routing design. In: Mathematical Methods in Economnics 2018, MatfyzPress, Praha, pp. 252–257 (2018)
15. Kvet, M.: Computational study of radial approach to public service system design with generalized utility. In: Proceedings of International Conference Digital Technologies 2014, Žilina, Slovakia, pp. 198–208 (2014)
16. Kvet, M., Janáček, J.: Spider network search strategy for p-location problems. In: CINTI 2020: IEEE 20th International Symposium on Computational Intelligence and Informatics, Budapest, pp. 49–54 (2020)
17. Marianov, V., Serra, D.: Location Problems in the Public sector, Facility location - Applications and theory (Z, Drezner, pp. 119–150. Springer, Berlin (2002). https://doi.org/10.1007/978-3-642-56082-8_4

18. Rybičková, A., Mocková, D., Teichmann, D.: Genetic algorithm for the continuous location-routing problem. Neural Netw. World **29**(3), 173–187 (2019)
19. Snyder, L.V., Daskin, M.S.: Reliability models for facility location. Exp. Fail. Cost Case, Tran. Sci. **39**(3), 400–416 (2005)
20. Zelinka, I.: SOMA-Self–organizing migrating algorithm. In: Davendra, D., Zelinka, I. (eds.) Self-Organizing Migrating Algorithm-Methodology and Implementation, Springer, pp. 3–49 (2016). https://doi.org/10.1007/978-3-319-28161-2

Extended Path-Relinking Method for p-location Problem

Jaroslav Janáček and Marek Kvet[⊠]

Faculty of Management Science and Informatics, University of Žilina, Univerzitná 8215/1, 010 26 Žilina, Slovakia
{jaroslav.janacek,marek.kvet}@fri.uniza.sk

Abstract. Most of the searching strategies based on path-relinking method usage are restricted by a drawback of the method. The drawback of the original path-relinking method consists in its way of processing the pair of input solutions. The path-relinking method applied to zero-one programming problems examines one of the shortest paths connecting the input solutions in the surface of a unit hypercube. This characteristic does not enable to examine any feasible solutions outside the sub-space determined by components, in which the input solutions differ. Within our research directed to heuristics for the public service system design problems, we suggested a new type of the path-relinking method, which is able to overcome the above-mentioned drawback. The novelty consists in determination of an infeasible solution of the p-location problem, which corresponds to a hypercube vertex with more than p-components, and in projection of a starting feasible solution in the set of the feasible solutions, which are the closest ones to the infeasible solution. The suggested path-relinking projective method was embedded into a simple one-to-all searching strategy and its efficiency dependent on infeasibility level of the infeasible solution was studied.

Keywords: Location problems · Heuristics · Path-relinking method extension

1 Introduction

The existence of human society has been always associated with decisions. Making more or less important decisions accompanies us in various areas of everyday life, although many times we are not even aware of it. We often encounter the requirement to find the optimal solution to a particular problem or to improve the current situation as much as possible. The main reasons for such rationalization include reducing costs and increasing efficiency. Choosing the right alternative from all solutions is not easy and involves a great deal of responsibility. The final decision may not affect only our personal lives, but also the lives of a certain group of people or even the whole society [16]. Another factor that needs to be taken into account when making a decision is the time aspect. The consequences of a decision can be very long. In this paper, we focus only on a strategic level of decision-making process. The time lag of strategic decisions is usually in the order of several years. Most often, these are large-scale investment projects, such

© Springer Nature Switzerland AG 2022
B. Dorronsoro et al. (Eds.): META 2021, CCIS 1541, pp. 70–79, 2022.
https://doi.org/10.1007/978-3-030-94216-8_6

as the construction of new companies, the location of distribution centers, or the design of various service systems. The research reported in this paper aims at applying the knowledge of Applied Informatics and programming in the location science, mainly to the healthcare segment [2, 4, 13].

The operation of the emergency medical service is one of the basic services by which the state protects its inhabitants and provides them with urgent care in critical situations [13]. The main role of each manager responsible for the efficiency of the service is to decide on the location of service centers. Centers, which can be, for example, warehouses, terminals, or specialized medical facilities, form the structure of the proposed system [16]. This structure plays an essential role in the efficiency of the system performance. Strategic decisions on the location of facilities so that the total costs are kept to be minimal or the service accessibility for patients to be as high as possible, represent a complex combinatorial problem, the solution of which can achieve significant savings or improve the quality of the service provided. Since the resources, which are to be located, are limited, the mathematical model used for the decision/making often follows the weighted p-median problem formulation [1, 7, 14]. To make the model more general, the concept of so-called generalized disutility has been introduced to consider also such requirements, which allow providing the service to a patient from more than one nearest located service centers. Even if this model extension makes the problem harder to be solved, it enables us to apply its results into a wider range of systems [9, 11, 15].

Wide range of practical applications of the weighted p-median problem not only in the medical sphere [13, 14, 16] has led to the creation of a large number of solving approaches, which include exact as well as heuristic and metaheuristic methods [1, 5, 6, 11, 19, 20].

Exact algorithms are based mostly on the branch and bound method. Sometimes, they may make use of the principles of duality. Their main disadvantage consists in their capacity limitation caused by commonly available universal optimization environments, to which the exact methods are embedded. Mentioned restriction does not usually allow us to solve problems of practical and real world size. On the other hand, there is a radial formulation of the problem [7, 14], which enables us to overcome this weakness. Other approach consists in developing a special software tool. Therefore, many Operations Research scientists and other authors focus mainly on heuristic and metaheuristic approaches [17, 18, 21].

Currently, the main attention is paid to various metaheuristic approaches, i.e. genetic algorithms, scatter search, path-relinking method and many others, the aim of which can be specified as a task of obtaining a good solution in acceptably short computational time. In this paper we report our research, which was aimed at extending the path-relinking method. This approach proved to be suitable mainly in the case of the generalized weighted p-median problem, in which the demands for service are assumed to occur randomly. It must be noted that the original path-relinking method inspects only the shortest path between two solutions. The scientific effort reported in this paper was aimed at suggesting such a version, which could project a starting solution into a feasible solution, which is the closest one to a given vertex of a unit hypercube regardless of its infeasibility. The suggested path-relinking projective method was embedded into a simple one-to-all search strategy and its efficiency depending on infeasibility level of the infeasible solution was studied.

2 Path-Relinking Method and Its Applications

The original path-relinking method was suggested to enable heuristic solution of the problems, which can be described by zero-one mathematical programming tools [8]. A general zero-one programming problem can be formulated by (1).

$$\min\{f(\mathbf{x}) : \mathbf{x} \in \mathbf{X} \subseteq \{0, 1\}^m\} \tag{1}$$

The idea of the method consists in searching one of the shortest paths connecting two input feasible solutions – vertices of an m-dimensional hypercube and returning the best-found-solution, which lies on the path. The hypercube vertices of the path correspond to m-dimensional vectors, components of which take values of one or zero. The sequential search along the shortest path is performed by a move from a currently occupied solution to a neighboring one, which differs from the occupied solution only in a value of one component. In addition, this component must belong to the set of components, which take different values in the vectors describing the input solutions. The original path-relinking method proceeds in accordance to the following algorithm applied to a pair $\mathbf{x}$, $\mathbf{y}$ of input solutions – m-dimensional zero-one vectors.

0. Define set D of components, in which $\mathbf{x}$ and $\mathbf{y}$ differ, i.e. $D = \{i = 1, \ldots, m: x_i \neq y_i\}$. Initialize $\mathbf{x}^{best}$ by $\mathbf{x}^{best} = argmin\{f(\mathbf{x}), f(\mathbf{y})\}$.
1. If $|D| > 1$ go to 2, otherwise go to 3.
2. Determine $d \in D$ by $d = argmin\{f(inv(\mathbf{x}, i)): i \in D\}$ and perform $\mathbf{x} = inv(\mathbf{x}, d)$, $D = D - \{d\}$. If $\mathbf{x} \in \mathbf{X}$ then update $\mathbf{x}^{best}$ by $\mathbf{x}^{best} = argmin\{f(\mathbf{x}^{best}), f(\mathbf{x})\}$.
3. Return $\mathbf{x}^{best}$ and terminate.

Comment: The operation $inv(\mathbf{x}, i)$ performed with m-dimensional zero-one vector $\mathbf{x}$ and subscript i from the domain $1, \ldots, m$ returns vector $\underline{x}$, components of which are defined as follows $\underline{x}_i = x_i$ for $i = 1, \ldots, m$, $i \neq d$ and $\underline{x}_d = 1 - x_d$.

The cardinality $|D|$ of the initial set D corresponds to the Hamming or Manhattan distance of the input solutions $\mathbf{x}$ and $\mathbf{y}$, and $|D|$-1 of inner vertices is the number of inner vertices on the shortest path connecting the input solutions in the surface of the m-dimensional unit hypercube. Efficiency of the path examination is obviously influenced by the number of feasible solutions inspected during the examination.

If a kind of p-location problem is considered, e.g. the weighted p-median problem or the emergency service system design problem with p service centers, then the set $\mathbf{X}$ of all feasible solutions is defined by (2).

$$\mathbf{X} = \left\{ \mathbf{x} \in \{0, 1\}^m : \sum_{i=1}^{m} x_i = p \right\} \tag{2}$$

Applying the above original version of the path-relinking method to the p-location problem, it can be found that at least every second vertex of the examined path will be inadmissible or infeasible solution. It means that the associated vector **x** will have less or more than p non-zero components. That is why, a more efficient version of the path-relinking method was suggested to solve problem (2). The new version avoids the weird vertices of the hypercube and inspects only feasible solutions of (2).

This adjusted path-relinking mod performs according to the following scheme.

0. Define sets D and E of components, in which take the value of one only in one of the input solutions **x** and **y**. $D = \{i= 1, ..., m: x_i = 1$ and $y_i = 0\}$ and $E = \{i= 1, ..., m: x_i = 0$ and $y_i = 1\}$. Initialize $\mathbf{x}^{best}$ by $\mathbf{x}^{best} = argmin\{f(\mathbf{x}), f(\mathbf{y})\}$.
1. If $|D| > 1$ go to 2, otherwise go to 3.
2. Determine $d \in D$ and $e \in E$ by $[d, e] = argmin\{f(swap(\mathbf{x}, i, j)): [i, j] \in D \times E\}$ and perform $\mathbf{x} = swap(\mathbf{x}, d, e)): D = D - \{d\}$, $E = E - \{e\}$, and update $\mathbf{x}^{best} = argmin\{f(\mathbf{x}^{best}), f(\mathbf{x})\}$. Go to 1.
3. Return $\mathbf{x}^{best}$ and terminate.

Comment: The operation $swap(\mathbf{x}, d, e)$ performed with m-dimensional zero-one vector **x** and subscripts d and e from the domain $1, ..., m$, for which $x_d = 1$ and $x_e = 0$ returns vector $\underline{\mathbf{x}}$, components of which are defined as follows $\underline{x}_i = x_i$ for $i = 1, ..., m$, $i \neq d$ and $i \neq e$. Furthermore $\underline{x}_d = 0$ and $\underline{x}_e = 1$.

The above-described path-relinking method proved to be an excellent tool when embedded into searching scheme of a discrete version of particle swarm optimization. Nevertheless, the domain of examined solutions stays restricted by the initial deployment of swarm particles and the system of shortest paths among them. To overcome this disadvantage of the method, we suggested an extended version of the path-relinking method described in the next section.

3 Concept of Projection and Path-Relinking Method Extension

The idea of extension is based on the m-dimensional unit hypercube geometry, where the set of feasible solutions (2) corresponds to a sub-set of the hypercube vertices, which lie in the intersection of the hypercube and a facet of the simplex determined by the constraint in (2).

Let us consider a vertex **v** of the hypercube, which does not belong to set of feasible solutions due to the number of its non-zero components exceeds the value of p. The vertex **v** induces a set $F(\mathbf{v})$ of feasible p-location problem solutions, which are the closest ones to the vertex **v** in terms of Hamming distance. As the vertex **v** has q non-zero components and $q > p$, the minimal Hamming distance equals to $q-p$.

Now, using the path-relinking principle, an input solution **x** will be projected to the set $F(\mathbf{v})$ and the best-found-solution of the shortest path from **x** and the set $F(\mathbf{v})$ will be an output of the procedure. Using the above introduced denotation, the extended path relinking method can be described by the following algorithm, input of which is a feasible solution **x** and an infeasible hypercube vertex **v** with q, $q > p$ components.

ExtendedPathRelinking (**x**, **v**)

0. Define $D = \{i= 1, ..., m: x_i = 1$ and $v_i = 0\}$ and $E = \{i= 1, ..., m: x_i = 0$ and $v_i = 1\}$. Initialize $\mathbf{x}^{best}$ by **x**.

1. If $|D| > 1$ go to 2, otherwise return $\mathbf{x}$ and terminate.
2. Determine $[d, e] \in D \times E$ by $[d, e] = argmin\{f(swap(\mathbf{x}, i, j)): [i, j] \in D \times E\}$ and update $\mathbf{x} = swap(\mathbf{x}, d, e)): D = D - \{d\}, E = E - \{e\}$, and $\mathbf{x}^{best} = argmin\{f(\mathbf{x}^{best}), f(\mathbf{x})\}$. Go to 1.

This extended path-relinking method can be employed in a simple version of a discrete particle swarm optimization algorithm [3, 21] with strategy one-to-all as follows.

Let $\mathbf{x}$ is a starting feasible solution of the solved p-location problem and V is a finite set of hypercube vertices, where each of them has more than p non-zero components. Then the searching strategy follows the next commands:

One-to-allSearch($\mathbf{x}$, V)

While $V \neq \emptyset$ do: Withdraw a $\mathbf{v}$ from V, update $V = V - \{\mathbf{v}\}$ and $\mathbf{x} = ExtendedPathRelinking(\mathbf{x}, \mathbf{v})$. If $V = \emptyset$, then terminate the search and return $\mathbf{x}$.

4 Numerical Experiments

To verify the extended path-relinking method, the medical emergency system design instances were used as benchmarks. The problem is formulated as a task to choose p centers out of the set of m possible center locations so that the objective function f is minimal. The collection of p chosen center locations can be described by an m-dimensional zero-one vector $\mathbf{x} \in \mathbf{X}$. Then, (3) can define the objective function f for the above-described problem. The formula expresses sum of mean distances from a system user j to the nearest available service center.

$$f(\mathbf{x}) = \sum_{j=1}^{n} b_j \sum_{k=1}^{r} q_k \min_k \{d_{ij} : i = 1, ..., m, x_i = 1\} \tag{3}$$

In the formulation (3), the operator $\min_k\{\}$ returns the k-th minimal value of the set $\{\}$. The function f is computed for n system users, where b_j denotes a number of user's demands, which are located at j and must be serviced from the nearest available service center. The time-distance between a user location j and a possible service center location i is denoted by symbol d_{ij}. The coefficients q_k, $k = 1, ..., r$ stand for probabilities that the k-th nearest service center is the closest available one. This problem description corresponds to the concept of emergency service system design, in which the system operates as a queuing system with p service lines. The system is characterized by a demand assignment strategy following the idea that a randomly emerged demand for service is assigned to the nearest service center only if the center is not occupied by an earlier demand. In the opposite case, the nearest non-occupied center provides the user with service [9, 11, 15].

Computational study reported in this paper was performed on benchmarks derived from real emergency medical service system implemented in eight self-governing regions of the Slovak Republic. These problem instances were used also in our previous research published in [10, 11]. The individual instances are denoted by the names of capitals of the particular regions, which are reported by abbreviations of the region denotations. The list of instances consists of Bratislava (BA), Banská Bystrica (BB), Košice (KE),

Nitra (NR), Prešov (PO), Trenčín (TN), Trnava (TT) and Žilina (ZA). The sizes of the individual benchmarks are m and p introduced above. Mentioned basic characteristics of all used benchmarks are reported in the left part of Table 1. The coefficients b_j used in the objective function (3) correspond to the number of inhabitants of individual communities rounded up to hundreds. The coefficients q_k for $k = 1...3$ of the generalized objective function (3) were set according to [12] at the values: $q_1 = 0{,}77063$, $q_2 = 0{,}16476$ and $q_3 = 1\text{-}q_1\text{-}q_2$. These values were obtained from a simulation model of existing emergency medical system in Slovakia. The middle part of the table consists the objective function value (3) of the optimal solution denoted by $OptSol$ together with the computational time in seconds denoted by CT [s], in which the optimal solution was obtained. The right part of Table 1 is devoted to the characteristics of the uniformly deployed sets as described in [10]. We report their cardinalities $|S|$ and minimal Hamming distance h. The uniformly deployed sets of solutions were used in the suggested solving heuristics as a source of feasible solutions of the problem. The process of uniformly deployed set construction and usage are reported in [10] and [11].

Table 1. Basic benchmarks characteristics, the optimal objective function values and uniformly deployed sets sizes

Region	m	p	Optimal solution		Uniformly deployed set	
			$OptSol$	CT [s]	$|S|$	h
BA	87	14	26650	0.35	23	2
BB	515	36	44752	10.57	172	3
KE	460	32	45588	7.58	60	2
NR	350	27	48940	19.21	83	2
PO	664	32	56704	76.53	232	2
TN	276	21	35275	4.04	137	2
TT	249	18	41338	2.79	212	2
ZA	315	29	42110	2.70	112	3

To construct the series V of the infeasible hypercube vertices for individual benchmarks, we ordered the corresponding uniformly deployed set S of p-location solutions according to objective function values. We used the best solution as the initial solution $\mathbf{x}$ and then, we grouped the remaining solutions to disjoint pairs, triples and quadruples. Each created group $\{\mathbf{x}^u: u = 1, ..., t\}$ of t solutions gave one vertex $\mathbf{v}$, components of which were determined according to $v_i = max\{x_i^u: u = 1, ..., t\}$. This way we solved four cases, where the first one did not use the infeasible vertices, but feasible solutions of S. The second case consisted of vertices obtained from pairs and thus $|V| = |S|/2$. In the third and fourth case the infeasible vertices were constructed from triples and quadruples respectively and cardinalities of V equaled to $|S|/t$ for $t = 3, 4$.

The main goal of this computational study is to verify the impact of the cardinality of V on the results measured by computational time in seconds and the solution accuracy.

Since the optimal objective function values of all studied benchmarks are available and published in [11], the quality of the resulting system design is here evaluated by *gap*, which expresses a relative difference of the obtained objective function value from the optimal one. Its value is reported in percentage, where the optimal objective function value was taken as the base. Obviously, we provide also the computational time *CT* in seconds.

To achieve the main goal of numerical experiments, a sufficient set of problems and uniformly deployed sets of solutions must be considered. To make the comparison relevant and robust enough, we followed from a very useful property of any uniformly deployed set of solutions. The mentioned useful feature consists in the fact that any arbitrary permutation of *m* locations subscripts brings a new set with the same parameters. This way, we were able to obtain ten different sets for each problem instance. The results are summarized in the following tables.

Table 2 contains the average results of ten instances solved for different uniformly deployed sets of solutions for each self-governing region.

For completeness, let us add the information that the numerical experiments were run on a PC equipped with the Intel® Core™ i7 3610QM 2.3 GHz processor and 8 GB of RAM. The algorithms were implemented in the Java language making use of the NetBeans IDE 8.2 environment.

Table 2. Average results of numerical experiments for the self-governing regions of Slovakia

Region	$|V| = |S|$		$|V| = |S|/2$		$|V| = |S|/3$		$|V| = |S|/4$	
	gap	CT	gap	CT	gap	CT	gap	CT
BA	1.19	0.19	1.81	0.14	2.32	0.09	2.97	0.07
BB	0.30	24.15	0.35	24.60	0.36	21.36	0.30	18.37
KE	0.37	13.90	0.37	13.55	0.36	11.63	0.55	9.68
NR	0.18	6.60	1.52	6.18	0.30	5.27	1.67	4.29
PO	0.45	16.17	0.59	16.86	4.91	15.39	4.89	13.96
TN	1.38	2.54	1.51	2.37	1.71	2.10	1.99	1.77
TT	0.23	1.54	0.10	1.43	0.14	1.21	0.12	1.00
ZA	0.07	7.00	0.05	6.55	0.05	5.47	0.05	4.47

The following Table 3 contains the results of the best run out of ten computations for each benchmark, in which the lowest value of the objective function (3) was achieved.

Table 3. The best results of numerical experiments for the self-governing regions of Slovakia (minimal objective function value of ten runs was taken into account)

Region	$|V| = |S|$		$|V| = |S|/2$		$|V| = |S|/3$		$|V| = |S|/4$	
	gap	CT	gap	CT	gap	CT	gap	CT
BA	0.00	0.19	0.65	0.14	0.68	0.08	1.12	0.06
BB	0.00	24.85	0.00	26.37	0.00	22.71	0.00	18.80
KE	0.00	14.36	0.00	13.27	0.00	11.59	0.00	9.40
NR	0.05	6.56	0.09	6.20	0.05	5.31	0.62	4.32
PO	0.03	15.93	0.12	16.77	3.57	15.17	3.57	13.73
TN	0.00	2.40	0.14	2.30	0.54	1.96	0.51	1.66
TT	0.00	1.57	0.00	1.43	0.00	1.24	0.00	1.07
ZA	0.00	7.05	0.00	6.55	0.00	5.57	0.00	4.95

5 Conclusions

The main purpose of this paper was to provide the readers with an effective heuristic method for solving middle and large instances of the weighted p-median problem, which finds its application in many different areas including medical sphere and many other subfields of location science. To make the solving approach applicable in a wider range, developed algorithm is able to cope with generalized objective function. The generalization consists in more service centers, which can provide the service to the system user and not only the nearest located center needs to be considered.

Suggested method is based on the former path-relinking method. The drawback of the original path-relinking method consists in its way of processing the pair of input solutions. Mentioned weakness was overcome and the reported computational results prove that most of the instances were solved either to optimality or the resulting solution was very near to the optimal one. The novelty of presented original method extension consists in determination of an infeasible solution of the p-location problem, which corresponds to a hypercube vertex with more than p-components, and in projection of a starting feasible solution in the set of the feasible solutions, which are the closest ones to the infeasible solution.

Based on performed numerical experiments we can conclude that we have constructed a very fast and effective heuristic approach to the p-location problems.

Future research in this scientific field could be concentrated on rules, which would enable to reduce the starting set of p-location problem solutions.

Acknowledgment. This work was supported by the research grants VEGA 1/0089/19 "Data analysis methods and decisions support tools for service systems supporting electric vehicles", VEGA 1/0689/19 "Optimal design and economically efficient charging infrastructure deployment for electric buses in public transportation of smart cities", and VEGA 1/0216/21 "Design of emergency systems with conflicting criteria using artificial intelligence tools". This work was supported by the Slovak Research and Development Agency under the Contract no. APVV-19-0441.

References

1. Avella, P., Sassano, A., Vasil'ev, I.: Computational study of large scale p-median problems. Math. Program. **109**, 89–114 (2007)
2. Current, J., Daskin, M., Schilling, D.: Discrete network location models. In: Drezner, Z., Hamacher, H.W. (eds.) Facility location, pp. 81–118. Springer Berlin Heidelberg, Berlin, Heidelberg (2002). https://doi.org/10.1007/978-3-642-56082-8_3
3. Davendra, D., Zelinka, I.: Self-Organizing Migrating Algorithm, Methodology and Implementation. Springer, Studies in Computational Intelligence, p. 289 (2016). https://doi.org/10.1007/978-3-319-28161-2
4. Doerner, K.F., Gutjahr, W.J., Hartl, R.F., Karall, M., Reimann, M.: Heuristic solution of an extended double-coverage ambulance location problem for Austria. CEJOR **13**(4), 325–340 (2005)
5. Drezner, T., Drezner, Z.: The gravity p-median model. Eur. J. Oper. Res. **179**, 1239–1251 (2007)
6. Elloumi, S., Labbé, M., Pochet, Y.: A new formulation and resolution method for the p-center problem. Informs J. Comput. **16**(1), 84–94 (2004)
7. García, S., Labbé, M., Marín, A.: Solving large p-median problems with a radius formulation. Informs J. Comput. **23**(4), 546–556 (2011)
8. Gendreau, M., Potvin, J.-Y. (eds.): Handbook of Metaheuristics. ISORMS, vol. 272. Springer, Cham (2019). https://doi.org/10.1007/978-3-319-91086-4
9. Janáček, J., Kvet, M.: Min-max optimization and the radial approach to the public service system design with generalized utility. In: Croatian Operational Research Review, Vol. 7, 1, pp. 49–61 (2016)
10. Janáček, J., Kvet, M.: Uniform deployment of the p-location problem solutions. In: Operations Research Proceedings 2019: Selected Papers of the Annual International Conference of the German Operations Research Society (GOR), Dresden, Germany, September 4–6, 2019, Springer, 2020, ISBN 978-3-030-48438-5, ISSN 0721-5924, pp. 315–321 (2019). https://doi.org/10.1007/978-3-030-48439-2_38
11. Janáček, J., Kvet, M.: Usage of uniformly deployed set for p-location min-sum problem with generalized disutility. In: SOR 2019 Proceedings, pp. 494–499 (2019)
12. Jankovič, P.: Calculating reduction coefficients for optimization of emergency service system using microscopic simulation model. In: 17th International Symposium on Computational Intelligence and Informatics, pp. 163–167 (2016)
13. Jánošíková, Ľ., Žarnay, M.: Location of emergency stations as the capacitated p-median problem. In: Quantitative Methods in Economics (Multiple Criteria Decision Making XVII). pp. 117–123 (2014)
14. Kozel, P. Orlíková, L., Pomp, M., Michalcová, Š.: Application of the p-median approach for a basic decomposition of a set of vertices to service vehicles routing design. In: Mathematical Methods in Economnics 2018, MatfyzPress, Praha, 2018, pp. 252–257 (2018)
15. Kvet, M.: Computational study of radial approach to public service system design with generalized utility. In: Proceedings of International Conference Digital Technologies 2014, Žilina, Slovakia, pp. 198–208 (2014)
16. Marianov, V., Serra, D.: Location problems in the public sector, Facility location - Applications and theory (Z, Drezner, pp. 119–150. Springer, Berlin (2002). https://doi.org/10.1007/978-3-642-56082-8_4

17. Rybičková, A., Burketová, A., Mocková, D.: Solution to the locating – routing problem using a genetic algorithm. In: SmaRTT Cities Symposiuum Prague (SCSP), pp. 1–6 (2016)
18. Rybičková, A., Mocková, D., Teichmann, D.: Genetic algorithm for the continuous location-routing problem. Neural Netw. World **29**(3), 173–187 (2019)
19. Sayah, D., Irnich, S.: A new compact formulation for the discrete p-dispersion problem. Eur. J. Oper. Res. **256**(1), 62–67 (2016)
20. Snyder, L.V., Daskin, M.S.: Reliability models for facility location. Expect. Fail. Cost Case, Transp. Sci. **39**(3), 400–416 (2005)
21. Zelinka, I.: SOMA-Self –organizing migrating algorithm. In: Davendra D., Zelinka, I., (eds.) Self-Organizing Migrating Algorithm-Methodology and Implementation, Springer, pp. 3–49 (2016). https://doi.org/10.1007/978-3-319-28161-2

The Construction of Uniformly Deployed Set of Feasible Solutions for the p-Location Problem

Peter Czimmermann$^{(\boxtimes)}$ (iD)

University of Žilina, Žilina, Slovakia
peter.czimmermann@fri.uniza.sk

Abstract. Feasible solutions of the p-location problems can be represented by n-bit binary words with exactly p ones. A set of selected solutions is called a t-uniformly deployed set, if the minimal Hamming distance between each pair of solutions is at least $2p - 2t$ for a given natural number t. The uniformly deployed sets can be used, due to their diversity, as starting population in evolutionary algorithms for p-location problems. In our contribution, we present a method for construction of appropriate t-uniformly deployed sets. The origins of this method trace back to the topological graph theory and we have adapted it to our purpose.

Keywords: Location problem · Uniformly deployed set · Voltage graph

1 Introduction

A lot of public service design problems are represented by weighted p-median and p-center problems. It is known that these problems belong to the family of hard computational problems [1]. Hence, various metaheuristics are used to solve them [2]. An important class of metaheuristics are evolutionary algorithms. These methods involve a set of starting feasible solutions. It is reasonable to suppose in location problems that this set has high diversity. Since the feasible solutions of p-location problems can be represented by n-bit binary words with exactly p ones, the diversity of the solutions can be measured by the Hamming distance [3]. A set of solutions with minimal Hamming distance $2p - 2t$ is called a t-uniformly deployed set ($t \in N$ is the maximum number of overlapping 1's in any two words).

Remark 1. We notice that the construction of the set, in which the Hamming distance between each pair of solutions is exactly $2p - 2t$, leads to the hard combinatorials problems with no fast algorithms to solve them. For example, we can point to the construction of difference sets or strongly regular graphs with given parameters.

© Springer Nature Switzerland AG 2022
B. Dorronsoro et al. (Eds.): META 2021, CCIS 1541, pp. 80–89, 2022.
https://doi.org/10.1007/978-3-030-94216-8_7

In this paper, we introduce a fast algorithm for the construction of t-uniformly deployed sets from voltage graphs (digraphs). The sets will be given by rows of an adjacency matrix of a digraph derived from a voltage digraph.

2 Definitions

In this section, we provide some important definitions of notions that are used later.

2.1 The p-Location Problem

The p-location problem is a task of locating p-centers at some of the n possible locations from the set I. It can be defined by (1), where the decision variable y_i gets the value one if a center is located at $i \in I$ and it gets zero otherwise.

$$min\{f(\boldsymbol{y}); y_i \in \{0,1\}, i \in I, \sum_{i \in I} y_i = p\} \tag{1}$$

Where $f(\boldsymbol{y})$ is an appropriate objective function. It is known that only few versions of $f(\boldsymbol{y})$ lead to the problems solvable in polynomial time [4,5].

2.2 Hamming Distance

Let two n-bit binary words $\boldsymbol{x} = (x_1, \ldots, x_n)$, $\boldsymbol{y} = (y_1, \ldots, y_n)$ be given. The Hamming distance of $\boldsymbol{x}$ and $\boldsymbol{y}$ is

$$H(\boldsymbol{x}, \boldsymbol{y}) = \sum_{i=1}^{n} |x_i - y_i|.$$

If the words $\boldsymbol{x}$ and $\boldsymbol{y}$ contain exactly p ones, then their Hamming distance is an even number

$$H(\boldsymbol{x}, \boldsymbol{y}) \in \{0, 2, 4, \ldots, 2p\}.$$

Let $\boldsymbol{x}$ and $\boldsymbol{y}$ represent two feasible solutions of the p-location problem and $H(\boldsymbol{x}, \boldsymbol{y}) = 2q$ (where $q \leq p$). The expression

$$\frac{2p - 2q}{2} = p - q = t$$

gives the number of locations contained in both solutions.

2.3 The t-Uniformly Deployed Sets

Let I_p be the set of all feasible solutions of a given p-location problem. Hence, I_p contains all n-bits binary words with exactly p ones. The t-uniformly deployed set is a subset $S \subseteq I_p$ such that the inequality $H(\boldsymbol{x}, \boldsymbol{y}) \geq 2p - 2t$ holds for each $\boldsymbol{x}, \boldsymbol{y} \in S$. It means that any two words $\boldsymbol{x}$ and $\boldsymbol{y}$ from S have at most t ones on the same positions [6].

2.4 Digraphs

For our needs, we use a more general definition of digraphs. A digraph D is a pair (V, E), where V is a non-empty set of vertices, and E is a set of directed edges. Every edge has exactly one starting vertex and one end vertex. Multiple edges and loops are also allowed. We say that e_1 and e_2 are multiple edges, if they have the same starting vertex and the same end vertex. We say that edge e is a loop, if it starts and ends at the same vertex. A monopole is a digraph that contains only one vertex, and all its edges are loops. Monopole with p edges is denoted by M_p. The outdegree of vertex u is the number of edges, which start at u. We say that vertex v is a successor of vertex u, if there is an edge from u to v. The adjacency matrix of a digraph is a square matrix $A = (a_{i,j})_{n \times n}$ such that $a_{i,j}$ represents the number of edges from i to j.

2.5 Groups and Modular Arithmetic

A group $(X, *)$ is a non-empty set X with binary operation $*$ defined on X such that

1. $\forall a, b \in X \quad a * b \in X$,
2. $\forall a, b, c \in X \quad (a * b) * c = a * (b * c)$,
3. $\exists e \in X$ such that $\forall a \in X \quad a * e = a = e * a$,
4. $\forall a \in X \quad \exists a^{-1} \in X$ such that $a * a^{-1} = e = a^{-1} * a$.

For example, integers Z with operation $+$ form the group $(Z, +)$. In modular arithmetic, there exists another important class of groups. The set of all remainders of division by k is denoted by Z_k. It means that

$$Z_k = \{0, 1, \ldots, k-1\}.$$

We can define addition $\oplus_k$ on Z_k by the expression

$$a \oplus_k b = mod(a + b, k)$$

where $a, b \in Z_k$ and $mod(x, y)$ is the remainder after dividing x by y. It is possible to show that $(Z_k, \oplus_k)$ is the group for any $k \in N$. Sometimes, we can omit the operation and parentheses, and the group $(X, *)$ can be denoted by X.

2.6 Voltage Digraphs

The construction of large graphs and digraphs from voltage graphs and digraphs is a method that was invented in topological graph theory [7]. This method was later used in the Degree/diameter problem [8–11]. Let a graph (digraph) $G = (V, E)$ and a group $(X, *)$ be given. If every edge $e \in E$ has assigned a value $\alpha(e) \in X$, then G is called voltage graph (digraph), and values $\alpha(e)$ on its edges are called voltages.

We say that $G_X = (V_X, E_X)$ is a graph (digraph) derived from G, if
1) its vertex set contains all ordered pairs from $V \times X$, (where the vertex $(u, i) \in$

$V \times X$ is denoted by u_i),

2) a pair of vertices $u_i, v_j \in V_X$ forms an edge from E_X if and only if there is an edge $e \in E$ from u to v in G such that $i * \alpha(e) = j$.

Example 1. We can consider the voltage digraph G and the group Z_3 in Fig. 1. The derived digraph G_{Z_3} can be seen in Fig. 2.

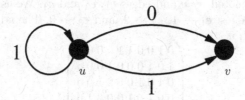

Fig. 1. Voltage digraph G.

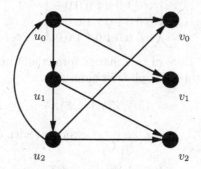

Fig. 2. Digraph derived from G.

3 The Construction of Uniformly Deployed Sets from Voltage Digraphs

In [8], there is shown the construction of the Hoffmann-Singleton graph from a voltage graph with two vertices and the group $Z_5 \times Z_5$. The Hoffmann-Singleton graph contains 50 vertices, each vertex has degree seven, each pair of adjacent vertices has no common neighbour, and each pair of non-adjacent vertices has exactly one common neighbour. It follows from these facts that any two rows of its matrix have the Hamming distance 12 or 14. Hence, the rows of this matrix form the 1-uniformly deployed set for $n = 50$ candidates and $p = 7$ locations. However, the Hoffmann-Singleton is a graph with many special properties. It is a strongly regular graph, a Moore graph, and a triangle-free graph. It is known that graphs with these properties occur rarely. This is the reason why we decided to construct digraphs with less constraints by this method. It is possible to show that a t-uniformly deployed set (n, p and t are given) can be obtained from adjacency matrix of a digraph on n vertices, in which every vertex has outdegree p and each pair of vertices has at most t common successors. We denote such digraphs by $D(n, p, \leq t)$. In this paper, we study possible constructions of $D(n, p, \leq t)$ from monopoles M_p and groups Z_n.

Example 2. We show the construction of 1-uniformly deployed set for $n = 10$ and $p = 3$ by this way. We start with monopole M_3 and group Z_{10}. Let the vertex of M_3 be denoted by v and edges e_1, e_2 and e_3. We assign the following voltages to these edges: $e_1 \to 1$, $e_2 \to 2$, and $e_3 \to 5$. The adjacency matrix of the derived digraph is

$$\begin{pmatrix} 0 1 0 0 1 1 0 0 0 0 \\ 1 0 1 0 0 0 1 0 0 0 \\ 0 1 0 1 0 0 0 1 0 0 \\ 0 0 1 0 1 0 0 0 1 0 \\ 1 0 0 1 0 0 0 0 0 1 \\ 1 0 0 0 0 0 0 1 1 0 \\ 0 1 0 0 0 0 0 0 1 1 \\ 0 0 1 0 0 1 0 0 0 1 \\ 0 0 0 1 0 1 1 0 0 0 \\ 0 0 0 0 1 0 1 1 0 0 \end{pmatrix}$$

We can check that the rows of this matrix form the 1-uniformly deployed set. All information about this digraph is in quintuple

$$(M_3, Z_{10}, 1, 2, 5).$$

In general, for M_p and Z_n, let v_i and v_j be vertices with common successor v_k. It means that the edges

$$(v_i, v_k), (v_j, v_k) \in E_{Z_n}.$$

Hence, there exist voltages $\alpha, \beta \in Z_n$ on edges of G such that $i \oplus_n \alpha = k$ and $j \oplus_n \beta = k$. From these equations, we obtain

$$j = i \oplus_n \alpha \oplus_n \overline{\beta},$$

where $\overline{\beta}$ is the inverse of β in Z_n. The list of all such vertices v_j, which have common successors with a given vertex v_i, can be obtained from the following table. We will call this table a range matrix for $(p+2)$-tuple

$$(M_p, Z_n, \alpha_1, \alpha_2, \ldots, \alpha_p).$$

$\oplus_n$	$\overline{\alpha_1}$	$\overline{\alpha_2}$	$\ldots$	$\overline{\alpha_p}$
α_1	0	$\alpha_1 \oplus_n \overline{\alpha_2}$	$\ldots$	$\alpha_1 \oplus_n \overline{\alpha_p}$
α_2	$\alpha_2 \oplus_n \overline{\alpha_1}$	0	$\ldots$	$\alpha_2 \oplus_n \overline{\alpha_p}$
$\vdots$	$\vdots$	$\vdots$	$\ddots$	$\vdots$
α_p	$\alpha_p \oplus_n \overline{\alpha_1}$	$\alpha_p \oplus_n \overline{\alpha_2}$	$\ldots$	0

The number of occurrences of value $\gamma \in Z_n$ in the range matrix is the answer to the question: how many common successors do the vertices v_i and $v_{i\oplus\gamma}$ have?

Example 3. The range matrix for $(M_3, Z_{10}, 1, 2, 5)$ is

$\oplus_{10}$	9	8	5
1	0	9	6
2	1	0	7
5	4	3	0

Each value from Z_{10}, except the zero, occurs in the range matrix at most once. It means that a vertex v_i has exactly one common successor with vertices $v_{i\oplus9}$, $v_{i\oplus6}$, $v_{i\oplus1}$, $v_{i\oplus7}$, $v_{i\oplus4}$, $v_{i\oplus3}$, and no common successor with vertices $v_{i\oplus}$, $v_{i\oplus}$, $v_{i\oplus}$, since the range matrix does not contain values 2, 5, and 8.

Example 4. A 1-uniformly deployed set for $n = 80$, $p = 8$ can be represented by 10-tuple

$$(M_8, Z_{80}, 1, 2, 4, 12, 21, 27, 34, 39).$$

The corresponding range matrix is

$\oplus_{80}$	79	78	76	68	59	53	46	41
1	0	79	77	69	60	54	47	42
2	1	0	78	70	61	55	48	43
4	3	2	0	72	63	57	50	45
12	11	10	8	0	71	65	58	53
21	20	19	17	9	0	74	67	62
27	26	25	23	15	6	0	73	68
34	33	32	30	22	13	7	0	75
39	38	37	35	27	18	12	5	0

Each value from Z_{80}, except the zero, occurs in the range matrix at most once. Hence, this 10-tuple represents a 1-uniformly deployed set.

4   How to Construct the Set of Voltages

The main computational problem is the construction of an appropriate set of voltages. We present this problem for the Bratislava Region, where we have 87 candidates for emergency stations and we need to choose 14 locations. We have $\binom{87}{14}$ possibilities how to do it and we also have the same number of possibilities for the set of voltages.

Hence, in this section, we present the algorithm for choosing the voltages to obtain the derived digraph with parameters $D(n, p, \leq t)$. We define for these purposes an increasing sequence $\{a_i\}_{i=1}^{\infty}$ of nonnegative integers. We will call it a t-sequence and it can be stated recursively:

1. $a_1 = 0$, $a_2 = 1$.

2. For $k > 2$, a_k is the minimum value such that for all $i \in \{1, \ldots, k-1\}$, the value $a_k - a_i$ occurs between values $a_j - a_i$ (where $1 \le i < j < k$) at most $t-1$ times. The first q members of t-sequence can be computed by the following algorithm:

Let $a_1 := 0$; $a_2 := 1$;
For $k = 3, \ldots, q$
        $x := a_{k-1} + 1$;
        $A_k := \{a_j - a_i; 1 \le i < j < k\}$;
        While $a_k = 0$
            $y := 1$;
            For $i = 1, \ldots, k-1$
                If $(x - a_i) \in A_k$ at most $t-1$ times
                    Then $y := y \cdot 1$;
                    Else $y := y \cdot 0$;
            If $y = 1$ Then $a_k = x$;
                Else $x := x + 1$;

Where A_k is multiset. Examples of the first $q = 15$ members of t-sequences for $t = 1, 2, 3, 4$ can be seen below:

$t = 1$ $0, 1, 3, 7, 12, 20, 30, 44, 65, 80, 96, 122, 147, 181, 203$
$t = 2$ $0, 1, 2, 4, 7, 11, 16, 22, 30, 38, 48, 61, 73, 86, 103$
$t = 3$ $0, 1, 2, 3, 5, 8, 12, 16, 21, 27, 33, 40, 48, 57, 71$
$t = 4$ $0, 1, 2, 3, 4, 6, 9, 13, 17, 22, 27, 33, 39, 46, 53$

If we want to construct a digraph $D(n, p, \le t)$ from monopole M_p, group Z_n and t-sequence for appropriate t, then we can use the following procedure:

1) If $a_p < n/2$, then the voltages on edges are the members of t-sequence. It follows from the properties of t-sequences that the digraph derived from $(M_p, Z_n, a_1, \ldots, a_p)$ is $D(n, p, \le t)$.
2) If $a_{p-x} < n/2$, and $a_{p-x+1} \ge n/2$ (for small $x \in N$, for example $x \in \{1, 2, 3, 4\}$), then the voltages are $\alpha_i = a_i$ for $i = 1, 2, \ldots, p - x$.
For $k \in \{p-x+1, \ldots, p\}$, $\alpha_k \in \{\alpha_{k-1} + 1, \ldots, n-1\}$ is the minimum value such that for all $i \in \{1, \ldots, k-1\}$, the values $\alpha_k \oplus_n \overline{\alpha_l}$ and $\alpha_i \oplus_n \overline{\alpha_k}$ occurring in the multiset

$$\{\alpha_i \oplus_n \overline{\alpha_j}; \forall i, j \text{ such that } 1 \le i, j \le k, i \ne j\}$$

at most t times.
3) If we still do not have p voltages, then we can increase $t := t + 1$ and repeat step 2 to complete the set of voltages.

Example 5. A 4-uniformly deployed set for the Bratislava Region can be constructed from

$$(M_{14}, Z_{87}, 0, 1, 2, 3, 4, 6, 9, 13, 17, 22, 27, 33, 39, 46)$$

where all voltages are computed by previous procedure from 4-sequence.

5   Limitations

What are the limits of parameters n, p, and t, when we construct t-uniformly deployed sets from M_p and Z_n? From the range matrix, we have inequality

$$p(p-1) \leq t(n-1),$$

where $p(p-1)$ is the number of non-zero elements in a range matrix and $n-1$ is the number of non-zero elements in Z_n. From this inequality, we have some upper and lower bounds for n, p, and t. Lower bounds for n computed from inequality

$$\frac{p(p-1)}{t} + 1 \leq n$$

can be seen in table

$p \backslash t$	1	2	3	4
10	91	46	31	24
20	381	191	128	96
30	871	436	291	219

Lower bounds for t computed from inequality

$$\frac{p(p-1)}{n-1} \leq t$$

can be seen in table

$p \backslash n$	100	200	300
10	1	1	1
20	4	2	2
30	9	5	3

From inequality

$$p^2 - p - t(n-1) \leq 0,$$

we have interval

$$p \in \langle \frac{1 - \sqrt{1 + 4t(n-1)}}{2}, \frac{1 + \sqrt{1 + 4t(n-1)}}{2} \rangle$$

and some upper bounds for p can be found in table

$n \backslash t$	1	2	3	4
100	10	14	17	20
200	14	20	24	28
300	17	24	30	35

6   Computational Results

Colleagues J. Janacek and M. Kvet tested the efficiency of using the UDS in various heuristics to solve the weighted p-median problem and its generalised version. Their results can be found in [3,6,12,13]. We tested Swap and Path-relinking heuristics for the weighted p-median problem in [14]. Some numerical results (for generalised p-median problem) can be seen in the following table (taken over from [15]), which shows the tests of the discrete self-organizing migrating algorithm (DSOMA) with and without UDS extension. Benchmarks for the tests are derived from the self-governing regions of Slovakia.

Regions	n	p	OptSol	$DSOMA_U$	$Time_U[s]$	DSOMA	Time[s]
BB	515	36	44752	44907	30.3	44923	30.1
KE	460	32	45588	45733	17.6	46099	17.5
NR	350	27	48940	48996	8.5	49986	8.3
PO	664	32	56704	56936	2.8	60476	20.5
TN	276	21	35275	35789	3.4	49260	3.2
TT	249	18	41338	41432	2.0	44090	2.0
ZA	315	29	42110	42140	8.7	42145	8.7

The columns of the table mean:
Regions - shortcuts of the self-governing regions of Slovakia (Bratislava region is omitted),
n - the number of candidates for placing an emergency station,
p - the number of emergency stations that need to be located,
Opt Sol - optimal value of the objective function for the generalised weighted p-median problem,
$DSOMA_U$ - values of the objective function obtained by DSOMA with UDS extension,
DSOMA - values of the objective function obtained by basic version of DSOMA,
TimeU, Time - computation time.

7   Conclusions

In our contribution, we introduce the construction of t-uniformly deployed sets from voltage graphs. We study possible constructions from monopoles with elements from Z_n as voltages. We also derive some limitations for these classes of graphs and groups. The constructions from more complicated voltage graphs and groups will follow in our next paper. The effect of using t-uniformly deployed sets in genetic algorithms is tested in [13,14], where the authors present its efficiency on real data from regions of Slovakia. The solutions that can be obtained by this method could have applications in real-life contexts, such as the location of emergency stations within certain environs.

Acknowledgements. This work was supported by the research grants VEGA 1/0342/18 Optimal dimensioning of service systems, VEGA 1/0089/19 Data analysis methods and decisions support tools for service systems supporting electric vehicles. This work was supported by the Slovak Research and Development Agency under the Contract no. APVV-19-0441.

References

1. Kariv, O., Hakimi, S.: An algorithmic approach to network location problems. SIAM J. Appl. Math. **37**, 513–560 (1979). Author, F.: Article title. Journal 2(5), 99–110 (2016)
2. Janacek, J., Janosikova, L., Buzna, L.: Optimized design of large-scale social welfare supporting systems on complex networks. In: Thai, M.T., Pardalos, P.M. (eds.) Handbook of Optimization in Complex Networks: Theory and Applications, Chapter 12. Springer, Boston (2012). https://doi.org/10.1007/978-1-4614-0754-6_12
3. Janacek, J., Kvet, M.: Usage of uniformly deployed set for p-Location min-sum problem with generalized disutility. In: SOR 2019: Proceedings of the 15th International Symposium on Operational Research, 494–499 (2019). ISBN 978-961-6165-55-6
4. Klein, C., Kincaid, R.: The discrete anti-p-center problem. Transp. Sci. **28**(1), (1994)
5. Czimmermann, P., Pesko, S.: A polynomial algorithm for a particular obnoxious facility location problem. In: SOR 2015: Proceedings of the 13th International Symposium on Operational Research, pp. 427–432 (2015)
6. Kvet, M., Janacek, J.: Population diversity maintenance using uniformly deployed set of p-Location problem solutions. In: SOR 2019: Proceedings of the 15th International Symposium on Operational Research, pp. 354–359 (2019). ISBN 978-961-616555-6
7. Gross, J., Tucker, T.: Topological Graph Theory. Dover Publications; Reprint edition (2012). ISBN 978-0486417417
8. Siagiova, J.: A note on the McKay-Miller-Siran graphs. J. Comb. Theor. Ser. B **81**, 205–208 (2001)
9. Vetrik, T.: An upper bound for graphs of diameter 3 and given degree obtained as Abelian lifts of dipoles. Discussiones Mathematicae Graph Theory **28**(1), 91–96 (2008)
10. Czimmermannova, O., Czimmermann, P.: On the number of voltages on walks in dipoles. J. Inf. Control Manag. Syst. **6**(2) (2008)
11. Loz, E.: Graphs of given degree and diameter obtained as abelian lifts of dipoles. Discrete Math. **309**, 3125–3130 (2009)
12. Janáček, J., Kvet, M.: Uniform deployment of the p-location problem solutions. In: Neufeld, J.S., Buscher, U., Lasch, R., Möst, D., Schönberger, J. (eds.) Operations Research Proceedings 2019. ORP, pp. 315–321. Springer, Cham (2020). https://doi.org/10.1007/978-3-030-48439-2_38
13. Janacek, J., Kvet, M.: Efficient incrementing heuristics for generalized p-Location problems. Central Eur. J. Oper. Res. **29**, 989–1000 (2020)
14. Janacek, J., Kvet, M., Czimmermann, P.: Kit of uniformly deployed sets for p-Location problems. Submitted in Annals of Operations Research (2021)
15. Kvet, M., Janacek, J.: Unpublished result (2021)

Continuous Optimization

PSOwp: Particle Swarm Optimisation Without Panopticon to Evaluate Private Social Choice

Vicenç Torra[1]($\boxtimes$)(iD) and Edgar Galván[2](iD)

[1] Department of Computing Science, Umeå University, Umeå, Sweden
vtorra@ieee.org
[2] Naturally Inspired Computation Research Group, Department of Computer Science, Maynooth University, Lero, Maynooth, Ireland
edgar.galvan@mu.ie

Abstract. In a recent paper we introduced differentially private random dictatorship as a private mechanism for social choice. Differentially private mechanisms are evaluated in terms of their utility and information loss. In the area of social choice it is not so straightforward to evaluate the utility of a mechanism. It is therefore difficult to evaluate a differentially private social choice mechanism. In this paper we propose to use a particle swarm optimization-like problem to evaluate our differentially private social choice method. Standard particle swarm optimization (PSO) can be seen in terms of a panopticon structure. That is, a structure in which there is a central entity that knows all of all. In PSO, there is a particle or agent that knows the best position achieved by any of the particles or agents. We propose here PSO without panopticon as a way to avoid an omniscient agent in the PSO system.

Then, we compare different social choice mechanisms for this PSO without panopticon, and we show that differentially private random dictatorship leads to good results.

1 Introduction

In our recent work [12], we studied random dictatorship [2,4] as a voting mechanism that satisfies differential privacy [3] under some conditions, and defined a variation of this method that is differentially private.

In data privacy [5,11,13] data protection mechanisms are often evaluated in terms of their utility. Data protection mechanisms based on secure multiparty computation are known to be good with respect to utility as they provide loss-less computation and do not make any perturbation on the output of the function. In contrast, data protection mechanisms that follow differential privacy [3] or k-anonymity [8] cause some information loss to the data or computation.

As a result of this, it is relevant to evaluate the utility of random dictatorship and of its differentially private version.

Nevertheless, the evaluation of the utility of a voting mechanism is an ill-defined problem. Voting mechanisms are usually evaluated in terms of their properties

B. Dorronsoro et al. (Eds.): META 2021, CCIS 1541, pp. 93–105, 2022.
https://doi.org/10.1007/978-3-030-94216-8_8

based on individual preferences. Examples of properties include (see e.g. [1,2,9]) Condorcet conditions, the independence of irrelevant alternatives, etc. These conditions are defined assuming that voters possess ordinal utility functions. That is, voters have an order on the alternatives (e.g., prefer alternative a_1 to alternative a_2). In contrast, it is not considered a numerical evaluation of each alternative (e.g., the utility of alternative a_1 is 0.8 and the one of alternative a_2 is 0.5).

It is known that a numerical utility model does not fit well with voting procedures. Observe that for any ordinal utility function (i.e., a_1 is preferred to a_2), there are infinitely many (numerical) utility functions compatible with the ordinal one. Moreover, if we consider (numerical) utility functions for each voter (i.e., a numerical value for each alternative as $u_i(a_1)$, $u_i(a_2)$, ... for voter i), the majority rule does not necessarily maximize the total utility. That is, if we define the social good of a selected alternative as the addition of voters' utility for this alternative (i.e., $\sum_i u_i(a)$ for selected alternative a), majority voting does not necessarily lead to the best option. The same applies to other social choice mechanisms.

In this paper we propose a federated learning [7] type of problem using particle swarm optimisation (PSO) [6] to evaluate private social choice mechanisms (as the one introduced in [12]). The goal is to find an optimal (aggregated) position that is the best for a set of agents. The problem is formulated as a particle swarm optimisation (PSO) problem [6] in which there is no omniscient agent with knowledge of the so-far best optimal position for all as it is the case for standard PSO. I.e., no panopticon, as we say. The best optimal position is obtained through successive voting in line with successive aggregations in federated learning.

The structure of this paper is as follows. In Sect. 2 we review probabilistic social choice and a differentially private version of it. In Sect. 3 we review particle swarm optimisation and introduce particle swarm optimisation without panopticon. This later approach is to avoid the system of particles omniscient on the best position of each particle. Section 4 discusses the evaluation of differentially private social choice in terms of particle swarm optimisation. The paper finishes with some conclusions and directions for future work.

2   Probabilistic Social Choice

Let I be a set of agents and A a set of alternatives. Let the goal be to select the preferred alternative for the set of agents. That is, the alternative that most of the agents prefer.

To formulate this problem we model agents preferences in terms of preference relations on the set of alternatives. That is, for agent i, the preference relation $\succeq_i$ is defined in terms of subsets of $A \times A$. In our context, we have only access to the best preferred option of an agent $i \in I$ and this is just its vote, an alternative $a \in A$. So, for all a', $a \succeq_i a'$ for this agent $i \in I$.

Plurality voting is to select the alternative that receives the most votes or preferences. In contrast, uniform random dictatorship proceeds as follows.

Method 1. *From [12]. This method selects an agent i in I according to a uniform distribution on I, and then uses $\succeq_i$ to select the most prefered alternative*

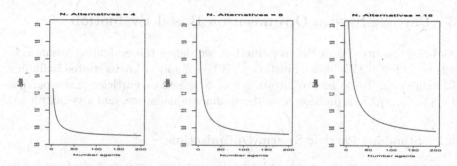

Fig. 1. Bounds for the ϵ parameter (y axis) in differentially private random dictatorship as defined in Method 2 using the results of Lemma 1 (read text). We have considered the case of 4, 8, and 16 alternatives (left, middle, and right graphs) and the number of agents ranging from 3 to 200 (x axis).

by agent i as outcome. That is, once i is selected from I, the method returns $a \in A$ such that $a \succeq_i a'$ for all $a' \in A$.

This approach can be equivalently implemented considering all alternatives, their frequency (votes), and then selecting one alternative using a probability distribution proportional to the frequency.

We defined in [12] two differentially private versions of random dictatorship. Their difference was on whether the voting was compulsory or optional. We give below the definition where voting is not compulsory but optional.

Method 2. *From [12], let $A = \{a_1, \ldots, a_m\}$ be the set of alternatives. Let I be the set of agents, and let $\succeq_i$ be the corresponding preference relations for $i \in I$ on the alternatives A. Then, enlarge I with a set of agents $I_0 = \{e_1, \ldots, e_m\}$ such that $\succ_i$ for $i \in I_0$ has as its prefered alternative the ith alternative in A.*

Then, apply uniform random dictatorship on $I \cup I_0$.

This voting procedure satisfies differential privacy for an appropriate parameter ϵ. The following lemma establishes bounds for the ϵ parameter.

Lemma 1. *From [12], differentially private random dictatorship as defined in Method 2 satisfies differential privacy for any*

$$\epsilon \geq \log \frac{2|I \cup I_0|}{|I \cup I_0| + 1}.$$

For a large number of agents, it is easy to see that we can compute a bound for ϵ. That is, $\epsilon > \log(2) = 0.6931$. Naturally, the more alternatives we have, the more agents we need to tend to this limit. Figure 1 represent the bound in Lemma 1 for 4, 8 and 16 alternatives. As we will describe later, we use in our experiments 8 alternatives. The corresponding figure shows the bound for a number of agents between 3 and 200.

When we can ensure that there is at least an agent for each alternative, we have bounds that do not depend on the alternative. Nevertheless, this is not necessarily the case in our scenario. See [12] for details.

3   Particle Swarm Optimisation Based Evaluation

As briefly mentioned in the introduction, we define the evaluation scenario in terms of PSO [6]. We have a function $f : \mathbb{R}^n \to \mathbb{R}$ and we are interested in finding its minimum. To do so, we have a set of S agents or particles. Each particle $i \in \{1, \ldots, S\}$ has a position p_i in the n dimensional space, and a velocity v_i.

3.1   Standard Particle Swarm Optimisation

In a standard particle swarm optimisation solution, each particle records the best position found so far. This is denoted by b_i. In addition, we keep track of the global best position found so far in the whole system. This is denoted by g.

The procedure iteratively computes a new position for each particle until a certain termination criteria is met. In each iteration, the best position is updated when necessary. More precisely, for the ith particle, we compute a new velocity:

$$v_i = \omega v_i + \phi_p r_p (b_i - p_i) + \phi_g r_g (g - p_i) \tag{1}$$

where ω is the inertia weight, ϕ_p and ϕ_g are acceleration coefficients one for the best position of the particle and the other for the best global position; and where r_p and r_g are random vectors following a uniform distribution in $[0, 1]$.

Then, we update the position of the ith particle as follows:

$$p_i = p_i + v_i.$$

When $f(p_i) < f(b_i)$ then we update the best position $b_i = p_i$, and if $f(p_i) < f(g)$ then we update the global best position $g = p_i$.

3.2   Particle Swarm Optimisation Without Panopticon

In PSO, the position of any agent or particle is public. Our scenario differs from the standard PSO scenario because we consider it private. Therefore, we cannot use the global best position g when computing a new position or velocity for any particle.

Instead, we consider an additional *system* particle that is led by all the particles. This *system* particle has its own position and velocity. We denote them by p_G and v_G, respectively. These position and velocity are public.

The position p_G is analogous to the aggregated model in federated learning. This position is based on agent's positions, and it is computed as an aggregation (using our social choice mechanisms) of previous and current information.

As the best global position g is not available, Eq. 1 cannot be used. Thus we compute the velocity of each particle in a slightly different way. As the system particle position is known and any particle can evaluate whether this position is better or not than its own, we update particles velocity taking advantage of this knowledge. Formally,

$$v_i = \begin{cases} \omega v_i + \phi_g r_g (p_G - p_i) & \text{if} f(p_G) < f(p_i) \\ \omega v_i + \phi_p r_p (b_i - p_i) & \text{otherwise.} \end{cases} \tag{2}$$

Equation 2 is similar to Eq. 1 but updating does not depend on g, and the updating rule depends on whether the ith particle is in a better position than system's one (i.e., $f(p_G) < f(p_i)$).

Updating of system's position needs to take into account that access to other particles' positions is not permitted. Our proposal is that particles can provide a direction where to lead the system particle. This direction plays the role of the velocity vector, but there are two main differences.

- One is that the direction is a vector but it does not have a magnitude.
- Another one is that not all directions are possible, but only a limited number of them.

We have these constraints because we consider that supplying an arbitrary direction or velocity is not feasible from a privacy perspective: the space of alternatives would be too large to protect (too many possible angles and magnitudes).

The number of directions n_d is a parameter of the system. In this work we only consider functions with two variables, so they are functions of the form $f : \mathbb{R}^2 \to \mathbb{R}$. Then, all n_d directions are on the plane.

When $n_d = 4$ it means that particles can vote for four directions and they correspond to the following direction vectors $(1,0)$, $(0,1)$, $(-1,0)$ and $(0,-1)$. In general, each possible direction $a = 0, \ldots, n_d$ corresponds to a different angle, all angles are equally spaced in $[0, 2\pi]$ and they are defined with respect to the $(1,0)$ vector. At a given time, each particle computes its angle with this vector $(1,0)$, say α_i, and then vote for the option $\lfloor \alpha_i \cdot n_d/(2\pi) \rfloor$ which is the nearest option to their own preferred angle.

Given a set of particles, from their votes for their preferred angle, we can select an angle using any social choice approach. In particular, we can use plurality voting (i.e., select the most frequent angle), random dictatorship, and differentially private random dictatorship for selecting an angle. This process leads to an angle α_G which can then be used to find a direction vector v_{α_G}. That is, v_{α_G} is the unit vector with angle α_G with the vector $(1,0)$. Once the vector is known, we update the global position as follows:

$$p_G = p_G + \omega_G v_{\alpha_G},$$

where ω_G is the inertia weight of the global position.

3.3 Analogy with Federated Learning

Our approach has similarities with the standard procedure in federated learning. Note that agents access p_G. So, we assume that this information is publicly available. This is similar to accessing the average model in federated learning. Then, the information that agents provide in our system, that is, direction, can be seen as the difference between the global model and the local model in federated learning. Our approach is more restrictive than in federated learning, as we are dealing with a context in which agents can only vote for a few options. This has, of course, advantages from a privacy point of view.

4 On the Evaluation of Differentially Private Social Choice

We have considered different scenarios in order to evaluate differentially private social choice. Different scenarios differ on the function to be minimized, the social choice procedure, and the parameters of the system. We discuss these elements below.

4.1 Functions

In order to evaluate our approach we have used the following functions in $\mathbb{R}^2$, selected from the review work by Sengupta et al. [10] on particle swarm optimisation. For each function we also include the range of the two variables (x_1, x_2).

We have selected functions in $\mathbb{R}^2$ because they provide a simple scenario with only a few voting options and compatible with the example in [12] of a cohort of drones guiding a ground vehicle. Drones vote continuously to guide the vehicle. The selected direction, landmark or position at any time is not so important. It is the overall set of decisions (the rough path) what influences the trajectory of the vehicle.

The functions we consider are the following ones.

- Quadratic function ($x_1, x_2 \in [-100.0, 100.0]$):

$$f_1(x_1, x_2) = x_1^2 + x_2^2$$

- Schwefel's problem 2.22 ($x_1, x_2 \in [-10.0, 10.0]$):

$$f_2(x_1, x_2) = |x_1| + |x_2| + |x_1| \cdot |x_2|$$

- Schwefel's problem 1.2 ($x_1, x_2 \in [-100.0, 100.0]$):

$$f_3(x_1, x_2) = x_1^2 + (x_1 + x_2)^2$$

- Generalized Rosenbrock's function ($x_1, x_2 \in [-2.0, 2.0]$):

$$f_4(x_1, x_2) = 100 * (x_2 - x_1 * x_1)^2 + (x_1 - 1)^2$$

- Generalized Schwefel's problem 2.26 ($x_1, x_2 \in [-500.0, 500.0]$):

$$f_5(x_1, x_2) = -x_1 sin(\sqrt{|x_1|}) - x_2 sin(\sqrt{|x_2|})$$

- Rastrigin's function ($x_1, x_2 \in [-5.12, 5.12]$):

$$f_6(x_1, x_2) = 2 \cdot 10 + x_1^2 - 10 cos(2x_1\pi) + x_2^2 - 10 cos(2x_2\pi)$$

- Ackley's function ($x_1, x_2 \in [-32.768, 32.768]$):

$$f_7(x_1, x_2) = -20e^{-0.2\sqrt{0.5(x_1^2 + x_2^2)}}$$
$$- e^{0.5cos(2x_1\pi) + cos(2x_2\pi)} + 20 + e$$

- Griewank function ($x_1, x_2 \in [-600.0, 600]$):

$$f_8(x_1, x_2) = 1 + (1/4000)(x_1^2 + x_2^2) - cos(x_1) * cos(x_2/\sqrt{2})$$

The optimal solutions for these problems correspond to a function equal to zero, except in the case of f_5 where the best solution corresponds to –12569.5.

4.2 Social Choice Procedures

The social choice procedures we have considered are:

- plurality voting,
- random dictatorship, and
- differentially private random dictatorship.

Social choice with plurality voting and random dictatorship are used as base line social choice procedures.

In addition, we have also implemented standard PSO. That is, there is an omniscient agent that observes all other agents and stores the best/optimal position found so far. This agent represents the guard in the panopticon.

Then, we consider for each of the three procedures above, two cases according to whether a particle votes or not. They are the following ones:

- A particle always votes;
- A particle only votes when its position is better than the best one, and in this case, decision to vote is based on a probability.

Thus, we have 7 different approaches: (i) PSO, (ii) plurality voting (PV), (iii) random dictatorship (RD), (iv) differentially private random dictatorship (DRD), and (v) plurality voting (bPV), (vi) random dictatorship (bRD), and (vii) differentially private random dictatorship (bDRD) only among those agents that have a position better than the global one.

4.3 Parameters

Our system is defined by the number of particles, the inertia weight ω, and the velocities ϕ_p, ϕ_g and the inertia weight ω_G. In addition, social choice procedures can have additional parameters corresponding to the probabilities related to when to vote. We assume that all particles use the same parameter's values.

We have used different sets of values for the parameters ω, ϕ_p, ϕ_g, and ω_G. They are the following ones.

- ω: 0.005, 0.001, 0.05, 0.1, 0.2, 0.4
- $\phi_p = \phi_g$: 0.01, 0.05, 0.1, 0.2, 0.5, 1.0, 2.0
- ω_G: 0.005, 0.01, 0.05, 0.1, 0.2, 0.4

We have used 50 particles and 1000 iterations to compare the results. That is, 1000 voting processes. A few additional examples that are shown in the figures have been considered with additional iterations (10000 iterations). We have used 8 voting options, corresponding to an angle of $2\pi \cdot a/8$ for $a = 0, \ldots, 7$ from direction $(1, 0)$. We have considered 30 executions, for each of the assignments considered.

4.4   Experiments

We evaluate the utility of differentially private random dictatorship using as test-bed the optimisation problems defined above and as a methodology to solve this problem the particle swarm optimisation without panopticon (as defined in Sect. 3.2) as well as using the social choice procedures in Sect. 4.2 with the parameters described in Sect. 4.3.

Our goal is to see if the differentially private random dictatorship has a comparable behaviour to the ones supposedly better of plurality voting and random dictatorship. To that end, we

– compare the solutions obtained using the three social choice procedures, and
– compare different parametrisations (when particles vote, parameters used).

In addition, we use PSO as the reference value. Nevertheless, as it keeps track of the best solution found so far, we expected PSO to outperform the other methods.

For each set of parameters considered, we have computed the mean of the optimal function found. That is, a mean of the values ($MeanF$) obtained for the 30 different executions. We have also recorded the minimum ($MinF$) obtained in these 30 executions. Table 1 displays optimal values of $MeanF$ and $MinF$ found for each function and each method: PSO – on the top row, right column; PV, RD, and DRD – middle row, from left to right; bPV, bRD, and $bDRD$ – bottom row, from left to right. Between brackets we display the parameters ω, $\phi_p = \phi_g$ and ω_G used to obtain the optimal solution.

PSO Vs. Social Choice Procedures. PSO is always better than any other social choice procedure. Except for problem f_5, PSO reaches always the minimum for the 30 executions. That is, except for f_5, both $meanF$ and $minF$ are always zero. Best PSO solutions are in most of the cases obtained with the parameters $\omega = 0.005$, $\phi_p = \phi_g = 2$, and $\omega_G = 0.005$.

As stated above, this is a natural consequence of PSO being omniscient and keeping track of the best positions found by any agent. Nevertheless, as we show below, the solutions of social choice procedures are also very good and equal in practice for most problems.

For problems f_1, f_2, f_3, f_4 and f_8 (see Sect. 4), the optimal values achieved for PSO are 0 and social choice procedures give solutions with values at least less than 0.009, often very close to zero, the global solution. In particular, for f_8 we find a solution with differentially private random dictatorship (DRD) with an objective function equal to $7.41 \cdot 10^{-9}$. See also solutions for f_1, f_2, f_3, f_4 in Table 1 that are virtually zero. Note that in the table we display both mean values ($MeanF$) and the best solution found ($MinF$).

For problem f_5 (with global minimum of -12569.5), PSO obtains a $meanF$ value of -668, and the best of the 30 executions leads to $minF$ equal to -837. Social choice solutions (both for $meanF$ and $minF$) have an optimal value of around -7. These are the worst results for both PSO and social choice procedures.

Table 1. Optimal values obtained for the functions $f_1, \ldots, f_8$ using PSO and the three social choice procedures (for both voting strategies: always voting, only voting if better than global optimum). For each function we display *MeanF* (top) and *MinF* (bottom) objective functions achieved. For each pair (function, *MeanF/MinF*), we have on the first line: Name of function, value displayed, result using PSO (and parameters ω, $\phi_p = \phi_g$ and ω_G of the optimal result). On the second line we have the results obtained using PV, RD, and DRD and on the third line the results obtained using bPV, bRD, and $bDRD$.

Function/PV/bPV	RD/bRD	PSO / DRD / bDRD
f_1	*MeanF*	0.0 (0.005 2.0 0.005)
$8.46 \cdot 10^{-6}$	$7.50 \cdot 10^{-6}$	$1.26 \cdot 10^{-5}$ (0.1 2.0 0.005)
$7.63 \cdot 10^{-6}$	$6.93 \cdot 10^{-6}$	$1.45 \cdot 10^{-5}$ (0.4 2.0 0.005)
f_1	*MinF*	0.0 (0.005 2.0 0.005)
$2.20 \cdot 10^{-6}$	$1.58 \cdot 10^{-6}$	$6.97 \cdot 10^{-8}$ (0.05 2.0 0.005)
$9.83 \cdot 10^{-8}$	$1.00 \cdot 10^{-6}$	$8.50 \cdot 10^{-8}$ (0.01 2.0 0.005)
f_2	*MeanF*	0.0 (0.005 2.0 0.005)
0.0033	0.0033	0.0037 (0.01 2.0 0.005)
0.0034	0.0032	0.0041 (0.2 2.0 0.005)
f_2	*MinF*	0.0 (0.005 2.0 0.005)
0.0017	0.0010	$9.70 \cdot 10^{-4}$ (0.05 1.0 0.01)
$4.95 \cdot 10^{-4}$	$6.59 \cdot 10^{-4}$	$2.50 \cdot 10^{-4}$ (0.2 0.5 0.005)
f_3	*MeanF*	0.0 (0.005 2.0 0.005)
$1.00 \cdot 10^{-5}$	$1.10 \cdot 10^{-5}$	$1.68 \cdot 10^{-5}$ (0.4 1.0 0.005)
$9.62 \cdot 10^{-6}$	$9.98 \cdot 10^{-6}$	$2.05 \cdot 10^{-5}$ (0.2 2.0 0.005)
f_3	*MinF*	0.0 (0.005 2.0 0.005)
$8.78 \cdot 10^{-8}$	$5.94 \cdot 10^{-7}$	$7.41 \cdot 10^{-8}$ (0.2 2.0 0.005)
$2.87 \cdot 10^{-7}$	$2.98 \cdot 10^{-7}$	$6.56 \cdot 10^{-7}$ (0.1 1.0 0.005)
f_4	*MeanF*	0.0 (0.2 2.0 0.005)
0.0667	0.0050	0.0070 (0.4 2.0 0.005)
0.0284	0.0027	0.0054 (0.2 2.0 0.005)
f_4	*MinF*	0.0 (0.005 2.0 0.005)
0.0063	$1.57 \cdot 10^{-4}$	$7.67 \cdot 10^{-5}$ (0.4 2.0 0.005)
$9.52 \cdot 10^{-4}$	$4.64 \cdot 10^{-6}$	$4.36 \cdot 10^{-5}$ (0.005 0.2 0.1)
f_5	*MeanF*	-668.00 (0.05 2.0 0.4)
-7.8904	-7.8904	-7.8903 (0.005 2.0 0.05)
-7.8905	-7.8905	-7.8905 (0.005 1.0 0.01)
f_5	*MinF*	-837.96 (0.005 2.0 0.005)
-7.89	-7.89	-7.89 (0.1 2.0 0.05)
-7.89	-7.89	-7.89 (0.01 0.05 0.01)
f_6	*MeanF*	0.0 (0.005 2.0 0.05)
6.96	0.83	1.18 (0.05 1.0 0.05)
6.18	1.02	1.19 (0.1 1.0 0.05)
f_6	*MinF*	0.0 (0.005 2.0 0.005)
$5.40 \cdot 10^{-4}$	$3.93 \cdot 10^{-4}$	$2.91 \cdot 10^{-5}$ (0.1 2.0 0.005)
$2.83 \cdot 10^{-4}$	$2.30 \cdot 10^{-4}$	$1.23 \cdot 10^{-4}$ (0.05 0.5 0.01)
f_7	*MeanF*	$4.44 \cdot 10^{-16}$ (0.005 2.0 0.005)
0.40	0.09	0.14 (0.4 0.1 0.05)
0.25	0.11	0.16 (0.05 0.1 0.05)
f_7	*MinF*	$4.44 \cdot 10^{-16}$ (0.005 2.0 0.005)
0.0043	0.0036	$6.18 \cdot 10^{-4}$ (0.05 2.0 0.005)
0.0011	0.0017	0.0019 (0.005 0.5 0.005)
f_8	*MeanF*	0.0 (0.005 2.0 0.005)
$3.08 \cdot 10^{-6}$	$2.50 \cdot 10^{-6}$	$5.16 \cdot 10^{-6}$ (0.2 2.0 0.005)
$2.56 \cdot 10^{-6}$	$2.64 \cdot 10^{-6}$	$4.30 \cdot 10^{-6}$ (0.005 2.0 0.005)
f_8	*MinF*	0.0 (0.005 2.0 0.005)
$8.05 \cdot 10^{-7}$	$5.11 \cdot 10^{-7}$	$7.41 \cdot 10^{-9}$ (0.2 1.0 0.005)
$2.62 \cdot 10^{-7}$	$5.59 \cdot 10^{-8}$	$1.61 \cdot 10^{-8}$ (0.4 0.05 0.01)

For the other problems, f_6, f_7 the best social choice solutions are $meanF$ = 0.83 and $minF$ = 2.91 · 10^{-5}, and $meanF$=0.0979 and $minF$=6.18 · 10^{-4}, respectively.

Social Choice Procedures and Differentially Private Random Dictatorship. Among the social choice procedures, for most of the problems the best solutions are either random dictatorship or differentially private random dictatorship. In some cases solutions are better by a factor of 10 or 100 to the one obtained with plurality voting. So, we can state that randomness is not an inconvenience but an advantage.

Only for f_3 the best solutions are obtained using plurality voting. However, in this case, the values achieved by the three social choice procedures are very similar. Observe in Table 1 that the values $meanF$ are 9.62 · 10^{-6} for bPV and 9.984 · 10^{-6} for bRD.

When we compare the case of agents always voting and the case of agents only voting when they have a position better than the global one, we have, as expected, that in most cases, results are better when only those agents with better positions vote.

With respect to parameters of the best solutions, there is more variety here than when using PSO. We can observe in the table that DPD and dDPD has most solutions with $\phi_g = \phi_p = 2.0$ but the best solutions for f_2 are with $\phi_g = \phi_p = 0.5$ and for f_3 are with $\phi_g = \phi_p = 1.0$.

Figure 2 shows the evolution of the system particle when differentially private random dictatorship is used. We can see that except for the problem f_6 the system particle tends to move to the optimal solution. For f_5 the optimal solution

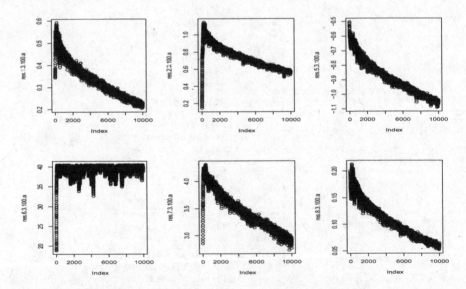

Fig. 2. Objective function for problems $f_1, f_2, f_5, f_6, f_7, f_8$ (right to left, top to bottom) listed above when differentially private random dictatorship is used. The number of alternatives considered is 8 and the number of particles is 100.

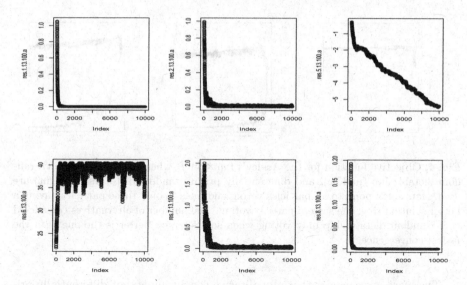

Fig. 3. Objective function for problems $f_1, f_2, f_5, f_6, f_7, f_8$ (right to left, top to bottom) when differentially private random dictatorship is used and when particles only vote if they know to be in a better position. The number of alternatives considered is 8 and the number of particles is 100.

found is far from optimal. This evolution is much faster when particles only vote when they know to be in a better position than the system particle. This is illustrated in Fig. 3.

To illustrate that the plurality vote is not always the best alternative, we show the results obtained for Ackley's function f_7. Figure 4 shows the results for the three social choice procedures: plurality rule (left), random dictatorship (middle), and differentially private random dictatorship according to Method 2 (right). Dots correspond to the case of all particles always voting, and lines to the case that only those particles with a better position vote. It can be clearly seen that the plurality rule is not best, and that voting only when a better position is found is clearly better. Recall that the optimal solution for this problem is when the function is exactly zero.

Summary. The results of our experiments can be summarised stating that

- standard PSO (with panopticon) is the most effective approach considered, but PSO without panopticon is also quite effective and some solutions have no significant difference,
- among social choice procedures implementing our variation of PSO (without panopticon), plurality voting is usually not the best option,
- particles voting only when their solution is a better approach than particles voting in all occasions, and, last but not least,
- differentially private random dictatorship can be seen as comparable to random dictatorship.

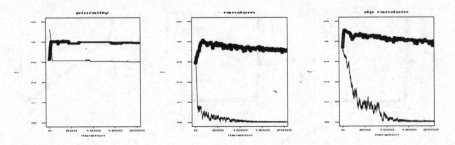

Fig. 4. Objective function for the Ackley's function f_7 when plurality rule (left), random dictatorship (middle), and differentially private random dictatorship (right) are used. Dots correspond to all particles voting and lines to only those particles having a better solution than the system particle voting. The number of alternatives considered is 8. Random dictatorship only voting when solutions are better is the one with the fastest convergence.

Therefore, we consider that a qualitative conclusion is that differentially private random dictatorship is a suitable approach to be used in this type of scenario.

5   Conclusions

This paper focuses on the evaluation of differentially private social choice and more particularly on differentially private random dictatorship. It is standard to evaluate data privacy mechanisms in terms of their utility or in terms of the loss they cause. Social choice mechanisms are not so straightforward to evaluate because the preferences or *opinions* of the agents are assumed to be expressed in ordinal terms. This is an important assumption. Our approach permits to evaluate social choice under these assumptions. We have proposed a scenario based on an objective function to optimise by a set of agents, and use a PSO-like procedure for obtaining the best solution through an iterative voting procedure. When numerical evaluations of the preferences exist (through e.g. utility functions), other mechanisms (as aggregation of utility functions) should be used.

We have shown that in our scenario, the results of differentially private random dictatorship are similar to those for random dictatorship, and usually better than those obtained with the plurality voting (i.e., selecting the most preferred option).

As future work we plan to study agents with different privacy requirements (e.g., privacy budgets) and how these different privacy requirements can affect the outcome of the system. Among the privacy options to consider, we have the case that agents want to refrain from voting (opt-out). Our experiment results used 50 particles in two dimensional problems, we will explore the case of larger number of particles and larger dimensions for the problem.

Acknowledgement. Discussions with Guillermo Navarro-Arribas are gratefully acknowledged. This study was partially funded by the Swedish Research Council (Vetenskapsrådet) (grant number VR 2016-03346), and by the Wallenberg AI, Autonomous Systems and Software Program (WASP) funded by the Knut and Alice Wallenberg Foundation.

References

1. Barberà, S.: Majority and positional voting in a probabilistic framework. Review of Economici Studies **42**(2), 379–389 (1979)
2. Brandt, F.: Rolling the dice: recent results in probabilistic social choice. In: Endriss, U. (ed.) Trends in Computational Choice, AI Access, pp. 3–26 (2017)
3. Dwork, C.: Differential privacy. In: Bugliesi, M., Preneel, B., Sassone, V., Wegener, I. (eds.) ICALP 2006. LNCS, vol. 4052, pp. 1–12. Springer, Heidelberg (2006). https://doi.org/10.1007/11787006_1
4. Gibbard, A.: Manipulation of schemes that mix voting with chance. Econometrica **45**(3), 665–681 (1977)
5. Hundepool, A., Domingo-Ferrer, J., Franconi, L., Giessing, S., Nordholt, E.S., Spicer, K., de Wolf, P.-P.: Statistical Disclosure Control. Wiley (2012)
6. Kennedy, J., Eberhart, R.C.: Particle swarm optimization. In: Proceedings IEEE International Conference Neural Networks, pp. 1942–1948 (2002)
7. Li, T., Sahu, A.K., Talwalkar, A., Smith, V.: Federated learning: challenges methods, and future directions, Arxiv (2019)
8. Samarati, P.: Protecting respondents' identities in microdata release. IEEE Trans. Knowl. Data Eng. **13**(6) 1010–1027 (2001)
9. Sen, A.: Collective Choice and Social Welfare. Penguin Books (2017)
10. Sengupta, S., Basak, S., Peters, R.A., II.: Particle swarm optimization: a survey of historical and recent developments with hybridization perspectives. Mach. Learn. Knowl. Extr. **1**, 157–191 (2018)
11. Torra, V.: Data Privacy: Foundations, New Developments and the Big Data Challenge. SBD, vol. 28. Springer, Cham (2017). https://doi.org/10.1007/978-3-319-57358-8
12. Torra, V.: Random dictatorship for privacy-preserving social choice. Int. J. Inf. Secur. **19**(5), 537–545 (2019). https://doi.org/10.1007/s10207-019-00474-7
13. Vaidya, J., Clifton, C., Zhu, M.: Privacy Preserving Data Mining. Springer, (2006). https://doi.org/10.1007/978-0-387-29489-6

Visualizing Crossover Rate Influence in Differential Evolution with Expected Fitness Improvement

Vladimir Stanovov[1]([✉]), Shakhnaz Akhmedova[2], and Eugene Semenkin[1]

[1] Siberian Federal University, 79 Svobodny pr.,
660041 Krasnoyarsk, Russian Federation
[2] Reshetnev Siberian State University of Science and Technology,
31 Krasnoyarsky rabochy pr., 660037 Krasnoyarsk, Russian Federation
`shahnaz@inbox.ru`

1 Introduction

The area of heuristic optimization methods, which includes evolutionary algorithms and biology-inspired methods, is currently under rapid development of due to the high efficiency of such approaches in various domains and availability of large computational resources. However, the proposal of new ideas is always strongly connected to the level of understanding of these algorithms' inner functioning principles, which is often not as high as desired. Because of this, the development of methods which allow better understanding of algorithms behaviour, for example, the influence of parameter values, would promote intuition of researches and lead to new ideas and directions of investigation.

One of the most popular evolutionary optimization techniques today is the differential evolution (DE) algorithm, originally proposed in [10]. The DE has shown its superior properties compared to other approaches in numerous competitions and found a large variety of real-life applications, which makes this method an interesting research topic. However, one of the disadvantages of DE is its high sensitivity to parameter values, such as scaling factor F and crossover rate Cr [4]. Better understanding of these parameters' influence is one of the most important directions of studies about DE.

In this paper the expected fitness improvement (EFI) metric is proposed to visualize the parameter search space of crossover rates of modern DE modification, NL-SHADE-RSP algorithm. The expected fitness improvement shows the possible improvement that could be achieved with different Cr values, highlighting the areas of interest at different stages of search process. Based on the EFI heatmap profiles, the conclusions about crossover rate importance are made for different benchmark scenarios, such as biased, shifted and rotated goal functions, taken from the Congress on Evolutionary Computation (CEC) 2021 competition for single-objective optimization. The performed experiments shows that efficient control strategies could be applied for NL-SHADE-RSP crossover rate change.

The rest of the paper is organized as follows: Sect. 2 provides the related work and describes DE basics, Sect. 3 contains the description of EFI metric

© Springer Nature Switzerland AG 2022
B. Dorronsoro et al. (Eds.): META 2021, CCIS 1541, pp. 106–123, 2022.
https://doi.org/10.1007/978-3-030-94216-8_9

calculation method, Sect. 4 contains the experimental setup and results, as well as their discussion, and Sect. 5 concludes the paper, outlining the possible directions of further studies.

2   Related Work: Differential Evolution

Differential Evolution or DE is the population based evolutionary algorithm for solving real-valued optimization problems firstly introduced by R. Storn and K. Price in [10]. This algorithm became one of the most popular among researchers due to its simplicity (it is easy to implement and has just three parameters, which will be discussed later) and high efficiency [7]. The differential evolution algorithm is based on the idea that to find the optimal solution only the difference vectors between candidate solutions should be used.

The basic DE approach has two main phases: the initialization and search conducted by mutation, crossover and selection operators. During the initialization a set (or population) of candidate solutions (also called individuals) $x_i = (x_{i,1}, x_{i,2}, ..., x_{i,D})$, $i = 1, ..., NP$, $j = 1, ..., D$, is randomly generated in the search space:

$$S = \{x_i \in R^D | x_i = (x_{i,1}, x_{i,2}, ..., x_{i,D}) : x_{i,j} \in [x_{lb,j}, x_{ub,j}]\} \tag{1}$$

using the uniform distribution with D being the dimensionality of that space and NP or population size is the first parameter of the DE algorithm.

After initialization individuals iteratively change their position in the search space with aim to find the best solution (optimum). For this purpose three operators are used: mutation, crossover and selection. The search process starts with mutation and in the original DE approach the *rand/1* mutation strategy was introduced:

$$v_{i,j} = x_{r1,j} + F \times (x_{r2,j} - x_{r3,j}), \tag{2}$$

where $x_{i,j}$ is the j-th coordinate of i-th individual, index i is different from indexes $r1$, $r2$ and $r3$, which are also mutually different. It should be noted that in this formula the second parameter of the DE algorithm, namely the scaling factor F, chosen from $[0, 2]$, is used. Mentioned parameter has to be adjusted for an optimization problem in hand.

After mutation the crossover operator is applied to mutant vectors v_i, $i = 1, ..., NP$. One of the most commonly used crossover operators is the binomial crossover, where each gene of the mutant vector v_i is exchanged with the corresponding gene of x_i with a uniformly distributed random number from $[0, 1]$ and additional condition:

$$u_{i,j} = \begin{cases} v_{i,j}, & \text{if } rand(0,1) < Cr \text{ or } j = jrand \\ x_{i,j}, & \text{otherwise} \end{cases}. \tag{3}$$

Here $Cr \in [0, 1]$ or crossover rate is the last parameter of the DE algorithm, while the *jrand* is a randomly chosen index from $[1, D]$. Thus, the genetic information of both parent-individual x_i as well as the mutant vector v_i are combined to generate the trial vector u_i. That additional condition is required to make sure that at least one coordinate of the trial vector u_i is taken from the mutant vector v_i, otherwise there is a chance that the trial vector and the parent individual will be the same, which will then lead to unnecessary calculations during the selection step.

The second crossover operator often used in DE is the exponential crossover, which performs crossover of adjacent components of the vector. In the exponential crossover first an integer n_1 is chosen randomly in range $[1, D]$ to act as a starting point for crossover, and then the second index n_2 indicating the number of components to be taken from the mutant vector is determined by incrementing n_2 with Cr probability. The exponential crossover is then performed using indexes n_1 and n_2 as follows:

$$u_{i,j} = \begin{cases} v_{i,j}, & \text{if } j \in [n_1, n_1 + n_2) \\ x_{i,j}, & \text{otherwise} \end{cases} \tag{4}$$

To keep individuals in the search space, namely each j-th coordinate of the i-th mutant vector in the interval $[x_{lb,j}, x_{ub,j}], j = 1, ..., D$, the midpoint target bound constraint handling method [1] was applied. In this method if the component of the obtained vector is greater than the upper boundary or smaller than the lower boundary, its parent x_i is used to set the new value for the mutant vector.

Finally, during selection either the trial vector u_i or the parent individual x_i is carried to the next iteration. It is done according to their fitness values, which are usually determined by calculating the objective function values: if the trial vector u_i is better or equal to the parent individual x_i in terms of fitness, then the i-th individual in the population is replaced. The selection step is performed in the following way:

$$x_i = \begin{cases} u_i, & \text{if } f(u_i) \le f(x_i) \\ x_i, & \text{if } f(u_i) > f(x_i) \end{cases} \tag{5}$$

Nowadays, there are a lot of modifications of the differential evolution approach developed for solving various optimization problems, including one- or multi-objective constrained or unconstrained optimization problems. Most of these modifications are focused on its parameters adjustment or proposing new mutation strategies [5]. The following several well-known mutation strategies are commonly applied to the DE algorithm: *rand/2*, *best/1*, *best/2*, *current-to-best/1* and *current-to-pbest/1*.

The last mutation strategy mentioned here, namely *current-to-pbest/1*, is to be of particular interest. It was introduced in the JADE algorithm [13] and later used in the SHADE algorithm [11] and also in its various modifications. The *current-to-pbest/1* mutation strategy works as follows:

$$v_{i,j} = x_{i,j} + F \times (x_{pbest,j} - x_{i,j}) + F \times (x_{r1,j} - x_{r2,j}), \tag{6}$$

where *pbest* is the index of one of the $pb * 100\%$ best individuals, different from i, $r1$ and $r2$. Thus, to use this mutation strategy the pb parameter should be chosen.

It was established that the scaling factor F as well as the crossover rate CR affect algorithm's efficiency and should be chosen carefully for specific optimization problems. Therefore, using them for mutation and crossover operators with fixed values may cause poor results. In the JADE algorithm [13] parameters F and Cr are adjusted automatically, to be more specific, firstly for each individual x_i at each iteration t, the crossover probability Cr_i is independently generated according to a normal distribution of mean μ_{Cr} and standard deviation 0.1; obtained value is then truncated to $[0, 1]$. In the same manner the mutation factor F_i for each individual x_i at each generation t is independently generated according to a Cauchy distribution with location parameter μ_F and scale parameter 0.1. If the obtained value $F_i \leq 0$ then it is generated again, and if $F_i \geq 1$ then it is set to 1.

Similar ideas were used in the SHADE algorithm [11], and its mechanism for parameter adaptation can be described as follows. The historical memory of H cells $(M_{F,h}, M_{Cr,h})$ is maintained, each containing a couple of F and Cr values (in the SHADE approach the memory size was set to $H = 5$ and the current memory index was denoted as h). Thus, for mutation and crossover operators new parameter values are sampled with Cauchy distribution $F = randc(M_{F,k}, 0.1)$, and normal distribution $Cr = randn(M_{CR,k}, 0.1)$, k is chosen in range $[1, D]$ for each candidate solution. Both obtained values are then truncated to $[0, 1]$ the same way as it is done in the JADE algorithm.

Additionally, two arrays S_F and S_{Cr} are generated: if there was an improvement in terms of the fitness value, then the corresponding values of parameters F and Cr as well as the fitness value difference Δf are stored in these arrays. They are used at the end of the iteration to update the memory cells with weighted Lehmer mean [3]:

$$mean_{wL} = \frac{\sum_{j=1}^{|S|} w_j S_j^2}{\sum_{j=1}^{|S|} w_j S_j}, \tag{7}$$

where $w_j = \frac{\Delta f_j}{\sum_{k=1}^{|S|} \Delta f_k}$, $\Delta f_j = |f(u_j) - f(x_j)|$ and S is either S_{Cr} or S_F.

And finally the new memory cell values are updated: $M_{F,k}^{t+1} = mean_{wL}(F)$, $M_{Cr,k}^{t+1} = mean_{wL}(CR)$, where t is the current iteration number.

The JADE and SHADE algorithms as well as their modifications (for example, the L-SHADE approach [12]) also use an external archive A, which size is usually equal to NP. Solutions replaced during the selection step are stored in that external archive. The archive A is empty during the initialization and it is filled as the algorithm works: if the newly generated candidate solution is better than the parent individual in terms of the fitness value, then the parent is saved in the archive. If the archive is full, the new individuals replace randomly selected ones. The individuals from the archive A are used during the mutation step, namely individuals used to calculate new coordinates can be randomly selected as from the population so from the external archive.

It should be noted that in the L-SHADE algorithm additionally the population size NP changes from iteration to iteration: the linear reduction strategy was proposed for the population size adaptation [12]. The population size NP is recalculated at the end of each generation, and the worst individuals in terms of fitness are eliminated. The population size is calculated with the linear function depending on current number of function evaluations:

$$NP_{g+1} = round(\frac{NP_{min} - NP_{max}}{NFE_{max}} NFE + NP_{max}), \qquad (8)$$

where $NP_{min} = 4$ and NP_{max} are the minimal and initial population sizes, NFE and NFE_{max} are the current and maximal number of function evaluations.

3   Expected Fitness Improvement Metric

Every optimization method mainly relies on the fitness values, as long as the goal is to minimize/maximize these values. The described parameter adaptation techniques are designed to adjust parameter values so that higher fitness improvements are achieved. So, the parameter setting is highly dependent not only on the fact of the improvement, but also on the improvement value, like in SHADE algorithm. The problem of setting the parameters represents an optimization problem itself, so a better understanding of this problem structure is highly desirable.

To perform the visualization of the possible fitness improvements at every step of the search process with different crossover rates Cr the Expected Fitness Improvement (EFI) metric is proposed. The EFI is based on the following idea: at every iteration where EFI should be calculated a large set of solutions is generated using mutation and crossover steps with different Cr values from a grid, and for every Cr the average improvement is measured. For example, to estimate the expected fitness improvements for the full range of crossover rates at a given generation g, the values of $Cr = 0, 0, Cr_{st}, ..., 1 - Cr_{st}, 1$ are tested, where Cr_{st} is the step size. The number of steps is defined as $NCr_{st} = \frac{1}{Cr_{st}}$. The result is an array EFI_{g,Cr_k}, $k = 0, ..., NCr_{st}$ for all $g = 1, ..., NG$ generations. The pseudocode of the EFI estimation for different Cr values is presented in Algorithm 1.

The EFI array containing the measured possible improvements could be visualized to estimate the distribution of promising crossover rate Cr values and the efficiency of parameter tuning technique used. However, there is a problem of values scale, which arises from the fact that at every next generation the average fitness improvements are gradually decreasing, i.e. if initially the EFI values could be around 10^{10} or even more, at the end of the search they could be around 10^{-10} or even exactly zero. To overcome this issue, the distribution of EFI values should be visualized at every generation separately.

The described EFI metric is applied to the NL-SHADE-RSP algorithm, developed for the CEC 2021 benchmark. It which contains several important improvements compared to the well-known L-SHADE, namely the non-linear

---

Algorithm 1. EFI computation

1: Initialize Differential Evolution
2: Set grid with Cr_{st}, NCr_{st}
3: Initialize matrix $EFI[NCr_{st}, NG] = 0$
4: Generation number $g = 0$
5: **while** $NFE < NFE_{max}$ **do**
6:    **for** $s = 0$ to NCr_{st} **do**
7:       Set $AImp = 0$
8:       **for** $i = 1$ to NP **do**
9:          Sample F value
10:          **if** $s == NCr_{st}$ **then**
11:             Sample Cr value
12:          **else**
13:             Set $Cr = s * Cr_{st}$
14:          **end if**
15:          Mutation
16:          Crossover
17:          **if** $f(u_i) < f(x_i)$ **then**
18:             $AImp = AImp + f(x_i) - f(u_i)$
19:             Save new solution
20:          **end if**
21:       **end for**
22:       $EFI[s, NG] = AImp/NP$
23:    **end for**
24:    $g = g + 1$
25:    Update algorithm specific parameters
26: **end while**
27: Return matrix $EFI[NCr_{st}, NG]$

---

population size reduction, adaptive archive usage, and modified historical memory size depending on the problem dimension. The population size in NL-SHADE is controlled in the following way:

$$NP_{g+1} = round((NP_{min} - NP_{max})NFE_r^{1-NFE_r} + NP_{max}), \qquad (9)$$

where $NFE_r = \frac{NFE}{NFE_{max}}$ is the ratio of current number of fitness evaluations. This population size control scheme was taken from the Adaptive Gaining-Sharing Knowledge (AGSK) algorithm [6], which implements the concept of knowledge exchange between experienced and non-experienced individuals in the population. The main operators and algorithm structure is still similar to DE, although there is a difference in trial vector generation - the algorithm generates them using two populations of mutant vectors. Although the paper does not describe it, according to the available source code, AGSK implements non-linear population size reduction presented above.

The NL-SHADE-RSP uses automatic tuning of archive usage probability, originated from the strategy adaptation implemented in IMODE algorithm [8]. The probability p_A of archive usage in the last index $r2$ in current-to-pbest strategy is initially set to 0.5, unless the archive is empty. It is then automatically tuned based on the number of usages n_A, which is incremented every time an offspring is generated using archive and sum of fitness improvements achieved with the archive Δf_A and without it Δf_P. The archive usage probability is recalculated at the end of each generation as follows:

$$p_A = \frac{\Delta f_A/n_A}{\Delta f_A/n_A + \Delta f_p/(1 - n_A)}. \tag{10}$$

After this the probability p_A is checked to be within [0.1, 0.9] by applying the following rule: $p_A = min(0.9, max(0.1, p_A))$, similar to the rule used in IMODE algorithm [8].

The pb value for current-to-pbest mutation in NL-SHADE-RSP is controlled in a similar manner to the jSO algorithm [2], with the initial pb_{max} set to 0.4 and the final $pb_{min} = 0.2$. The same linear reduction of pb parameter is used, allowing wider search at the beginning and better convergence at the end.

The NL-SHADE-RSP algorithm used both exponential and binomial crossovers with equal probability, and the type of crossover to be used was randomly chosen for each individual.

In addition to the proposed EFI metric, the pairwise distance distribution is analyzed in this study. For this purpose, the Euclidean distance between all individuals in the population is estimated, and the histogram of distances is built at every generation separately. The distributions of EFI and distances for several scenarios of DE optimization are presented in the next section.

4   Experimental Setup and Results

The experiments in this study were performed using the benchmark functions presented for the Congress on Evolutionary Computation 2021 competition on single-objective optimization [9] because this framework considers eight cases of the same goal functions, i.e. basic, biased, shifted, rotated functions, and combinations of these modifications, e.g. biased, shifted and rotated at the same time. The set of functions contained 10 functions, which should be tested with the optimization method across dimensions 10 and 20. The maximum number of function evaluations $maxFE$ is set to 2×10^5 and 10^6 for $10D$ and $20D$ functions respectively.

For every function and every benchmark type the step size for checked crossover rates was set to 0.01, i.e. there were 100 crossover rates tested from 0 to 0.99. The initial population size was set to $30D$, as this appeared to be a reasonable setting in previous studies. The memory size was set to $20D$. The algorithm for EFI estimation was implemented in C++ and compiled with GCC under Linux Ubuntu 20.04, and the post-processing was performed using Python 3.6 and matplotlib library. The EFI arrays, as well as distance histograms are visualized as heatmap profiles. In addition, the best, average, worst fitness values, average EFI, average distance and average of parameter values in memory cells M_{Cr} are shown in the figures. As long as the population size was constantly changing and the number of generations spent at the beginning of the search and at the end of the search to evaluate the same number of solutions is different, the EFI was calculated not every generation, but every $0.002\frac{NFE}{maxFE}$ function evaluations, resulting in 500 iterations.

The first experiment was performed for the bent cigar function (F1) without any modifications, $10D$, the EFI heatmap is shown in Fig. 1. Better values are shown in yellow.

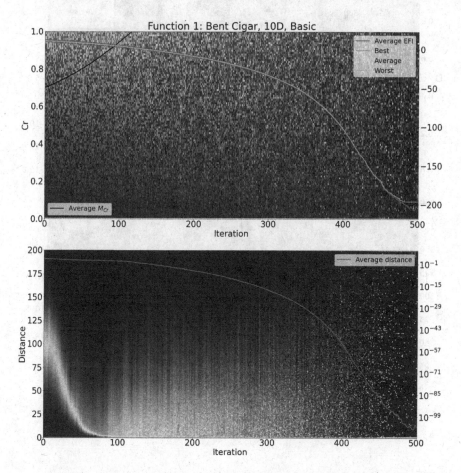

Fig. 1. EFI heatmap and distance histogram profiles, F1, basic benchmark, 10D

Figure 1 shows that for the relatively simple bent cigar function, where the achieved function values are around 10^{-200}, during most of the search process larger crossover rates were dominating, i.e. the expected improvement for Cr values was larger when $Cr > 0.5$ than for $Cr < 0.5$. This due to the fact that bent cigar is a non-separable function, i.e. it requires steps along more then one axis to perform the search. The distance histogram profile shows that there are no groups of points, which indicates that there is only one optimum. The average crossover rate in the memory cells quickly converges to 1, which seems to be a valid strategy in this case.

Figure 2 shows the results for shifted Schwefel's function, 10D.

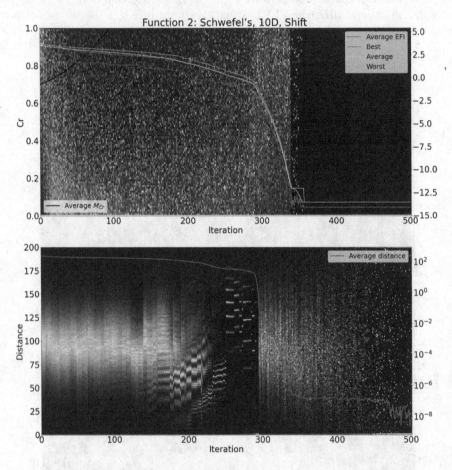

Fig. 2. EFI heatmap and distance histogram profiles, F2, shifted, 10D

The Schwefel's function has multiple local optima, and in non-rotated case could be efficiently solved by one-variable-at-a-time strategy, which is clearly seen on the EFI heatmap, where smaller Cr values perform better, as larger Cr lead to more "risky" moves in the search space, which not always result in efficient search. However, this is true only for the main part of the search process, i.e. from iteration 10 to iteration 300. During the first 10 iterations large Cr values are better, probably because in this period the initial exploration of the search space happens, capturing the most interesting areas to exploit later. Similar to this, the final convergence, which happens at around generation 300, and also requires larger Cr values, as at this moment it is important to find the optimum in a bowl-like landscape, so diagonal steps could be helpful. Also, the pairwise distance histogram shows that the algorithm identifies multiple local optima, and then deletes some of them thanks to the population size reduction. Despite the fact that smaller Cr are better, the memory cells values are dragged up to 1 due to the biased parameter adaptation with Lehmer mean.

Figure 3 shows the results for the same Schwefel's function, $10D$, but for the rotated case.

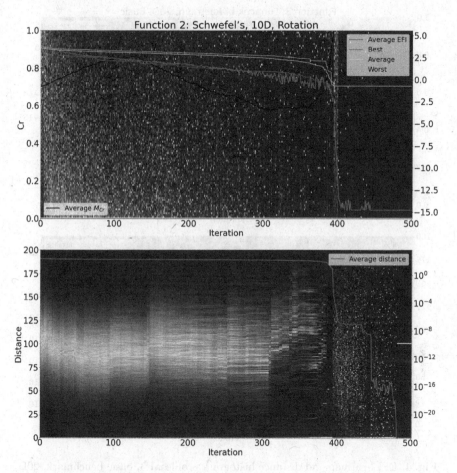

Fig. 3. EFI heatmap and distance histogram profiles, F2, rotated, 10D

In the rotated case the EFI heatmap changes for the Schwefel's function, but not in the way which could be expected. The first 10 iterations are almost the same, however, later the search efficiency drops, because it is difficult for the algorithm to tackle the function landscape. Although the function is rotated, large Cr values do not lead to significant improvements, while small Cr lead to some improvements, which makes the algorithm move the memory cells values towards zero. It could have been expected that for non-rotated functions smaller Cr would be more efficient, same as larger Cr for rotated, but the EFI in this cases shows that the opposite happens. In the rotated case the algorithm stopped without reaching the global optimum.

Figure 4 demonstrates the EFI heatmap profiles for the next function, Lunacek bi-Rastrigin, which has two large areas of attraction.

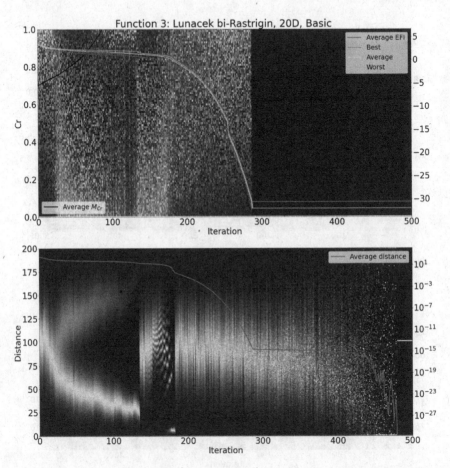

Fig. 4. EFI heatmap and distance histogram profiles, F3, basic benchmark, 20D

Figure 4 represents a particular interest, as here there are several switches between small and large Cr values being better for the function improvements. At the initial stage for the first around 20 generations the large Cr are dominating, and at the distance histogram it is clearly seen that all points have almost similar distances. After this period, the population is split in two groups, reach rushing towards one of the areas of attraction, and at this period smaller Cr

would result in more larger improvements - this could be because of the convergence to local optima, which is easily done by one-variable-at-a-time manner for non-rotated Rastrigin function. However, at the end of this period, at around iteration 100, large Cr start dominating again, when two parts of the population are far from each other. After the population size reduction cancels one of them out, small Cr values are better again, and finally, when the global optimum is almost found, Cr close to 1 are good again, allowing fast convergence to the optimum. These several switches demonstrate that even for such relatively simple problems the behaviour of Cr values could be quite complicated.

It is important to notice here how the average improvement, i.e. average EFI is related to the Cr switches and average fitness values in the population. When the red line, average EFI is above the yellow (average fitness) and green (worst fitness), $Cr > 0.5$ are dominating, and vice versa. This it true for all periods except the one around iteration 100, where all three lines are close to each other. Similar behaviour could be observed on all previous figures: if average EFI is closer or even larger than worst, then large Cr are better, and when average EFI is close to best fitness, smaller Cr appear to deliver more improvements. The mechanism behind such dependence remains unclear.

Figure 5 considers one of the more complicated cases, hybrid function with bias, shift and rotation applied altogether.

In the case shown on Fig. 5 there are two clearly seen stages of search: initial convergence and exploration with $Cr > 0.8$ being the best choice, and the more difficult and inefficient search, where smaller Cr allow better improvements. Same as for previous functions, the switch between these two stages, happening after iteration 100, coincides with the average EFI curve hitting worse and average fitness curves. The averaged memory cells M_{Cr} values in this case still keep the Cr close to 0.9, although the EFI shows that this is the region of smallest efficiency. This could be one of the reasons of algorithms low efficiency for this function. It is important to mention, that although smaller Cr are more efficient at the middle of the search process, this does not meant that the function is separable, is actually shows that performing the search along the axis at this stage would bring more benefits.

Considering the discovered dependency between the crossover rate, current fitness and average fitness improvement, a simple parameter control scheme was tested:

$$Cr = \begin{cases} 1, & \text{if } \frac{1}{|S|}\sum_{j=1}^{|S|} \Delta f_j > \frac{1}{NInds}\sum_{i=0}^{NInds} f_i \\ 0, & \text{otherwise} \end{cases}. \tag{11}$$

The crossover rate was updated every generation. In this case the EFI calculation was switched off. The comparison between NL-SHADE-RSP with and without fitness-based crossover rate control is presented in Table 1. The Mann-Whitney statistical test with significance level $p = 0.01$ is used for comparison, with the number of wins (+), ties (=) and losses (-) over 10 functions for every benchmark set and both $10D$ and $20D$.

Table 1. Mann-Whitney tests of NL-SHADE-RSP against modified with crossover control

Benchmark (code)	10D	20D
Basic (000)	1+/9=/0−	2+/8=/0−
Bias (100)	1+/9=/0−	3+/7=/0−
Shift (010)	1+/9=/0−	2+/8=/0−
Rotation (001)	1+/8=/1−	1+/8=/1−
Bias, Shift (110)	1+/9=/0−	2+/8=/0−
Bias, Rotation (101)	0+/9=/1−	2+/7=/1−
Shift, Rotation (011)	0+/8=/2−	2+/7=/1−
Bias, Shift, Rotation (111)	0+/8=/2−	2+/8=/0−
Total	5+/69=/6−	16+/61=/3−

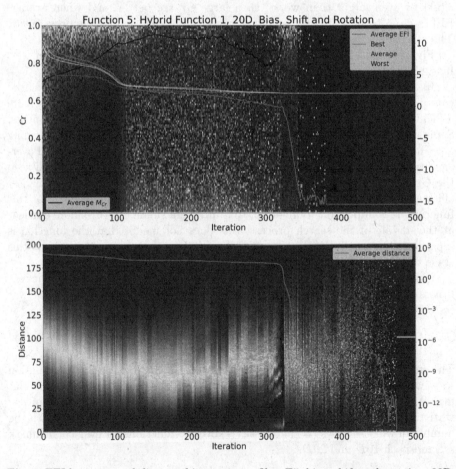

Fig. 5. EFI heatmap and distance histogram profiles, F5, bias, shift and rotation, 20D

The results in Table 1 show that such relatively simple control strategy could be competitive, or even better than the success-history based adaptation used in L-SHADE based algorithms. The improvements were observed for function 4 (Expanded Rosenbrock plus Rastrigin) for $10D$ and losses were for functions 5 and 6, i.e. hybrid functions. For 20 the wins were for functions 3 (Lunacek bi-Rastrigin), 4 and 6, while few performance deteriorations were for functions 8 and 9, i.e. composition functions. The improvements were mainly found at the end of the search, while for most of the computational resource both modified and non-modified NL-SHADE-RSP had similar performance - this is mainly due to the fact that Cr has much less influence on the algorithm performance, then, for example, scaling factor F.

For better understanding of the reasons of losses against success-history adaptation, in Fig. 6 the EFI heatmap profile is provided for F6, hybrid function 2, for the biased, shifted and rotated case.

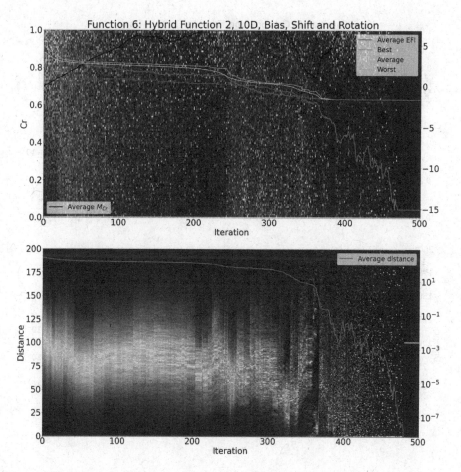

Fig. 6. EFI heatmap and distance histogram profiles, F6, bias, shift and rotation, 10D

In Fig. 6 it can be seen that the search process is relatively slow, but still taking place for the most of the time, i.e. there is no stagnation until iteration 320, and the crossover rate in the memory cells M_{Cr} is set to one, while the EFI heatmap profile indicates that smaller Cr are better. Considering the rule for Cr presented above, it would switch Cr to 0 in some cases, allowing to gain faster improvements, but probably leading the search away from the path where better final fitness values could have been achieved. This could be one of the reasons why the parameter adaptation has shown better results in this case. Another reason is that as long as in Eq. 16 the average improvements are calculated based on the few improvements available at current generation, the resulting value could be different from that calculated in EFI, as the last is averaged over a full range of Cr values. In other words, the average possible improvement without EFI calculation is biased, as long as it is estimated only with $Cr = 0$ or $Cr = 1$, depending on the results of previous generation.

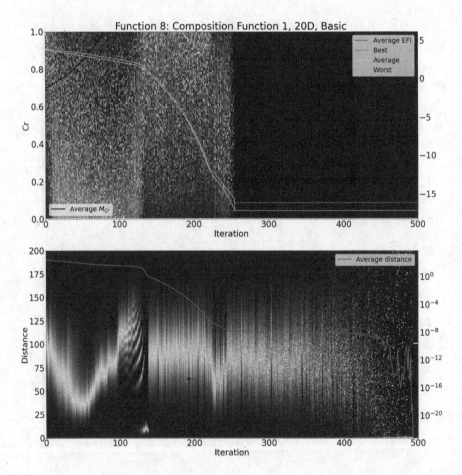

Fig. 7. EFI heatmap and distance histogram profiles, F8, basic, 20D

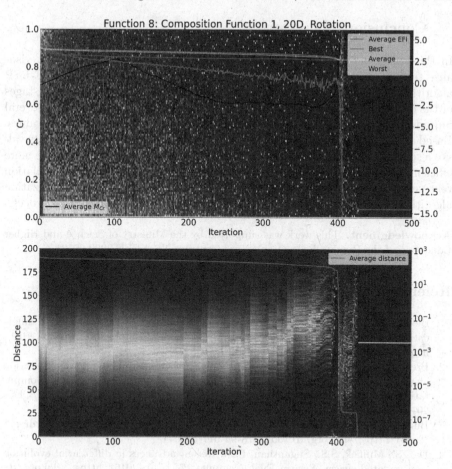

Fig. 8. EFI heatmap and distance histogram profiles, F8, rotation, 20D

Figures 7 and 8 provide the EFI heatmap profiles for composition function 1 (F8), for basic and rotated benchmark cases, 20D.ge

The contrast between the EFI heatmap profiles and distance histogram profiles demonstrate how different could be the search on the same function but with rotation. Obviously, currently available efficient DE variants do not have the rotational invariance property, and it cannot be achieved with the standard crossover strategies. Thus, some new crossover techniques should be developed which would be able to handle the rotated functions as good as non-rotated, for example with automatically de-rotating crossovers, which is yet to be developed.

5   Conclusion

In this study the expected fitness improvement metric was proposed to visualize the parameter search space of the crossover rate of the NL-SHADE-RSP algorithm, allowing to indicate more promising regions for Cr at different stages of the optimization process. The EFI heatmap profiles allowed revealing several important properties, such as switching behaviour of more promising Cr values. Based on the dependence between the improvements and the average fitness values a simple parameter control strategy is proposed, which was shown to be more efficient the standard parameter adaptation in some cases. The EFI calculation represents a general framework, which could be applied to other optimization algorithms to analyze different parameters' dynamics during the search process.

Acknowledgment. This work was supported by the Ministry of Science and Higher Education of the Russian Federation (State Contract No. FEFE-2020-0013).

References

1. Biedrzycki, R., Arabas, J., Jagodziński, D.: Bound constraints handling in differential evolution: an experimental study. Swarm Evol. Comput. **50**, 100453 (2019). https://doi.org/10.1016/j.swevo.2018.10.004
2. Brest, J., Maučec, M.S., Boškovic, B.: Single objective real-parameter optimization algorithm jSO. In: Proceedings of the IEEE Congress on Evolutionary Computation, pp. 1311–1318. IEEE Press (2017). https://doi.org/doi.org/10.1109/CEC.2017. 7969456
3. Bullen, P.S.: Handbook of Means and Their Inequalities. Springer, Dordrecht (2003). https://doi.org/10.1007/978-94-017-0399-4
4. Das, S., Mullick, S.S., Suganthan, P.N.: Recent advances in differential evolution - an updated survey. Swarm Evol. Comput. **27**, 1–30 (2016). https://doi.org/10.1016/j.swevo.2016.01.004
5. Das, S., Suganthan, P.N.: Differential evolution: a survey of the state-of- the-art. IEEE Trans. Evol. Comput. **15**(1), 4–31 (2011). https://doi.org/10.1109/TEVC.2010.2059031
6. Mohamed, A.W., Hadi, A.A., Mohamed, A., Awad Noor, H.: Evaluating the performance of adaptive GainingSharing knowledge based algorithm on CEC 2020 benchmark problems. In: Proceedings of the IEEE Congress on Evolutionary Computation, CEC, pp. 1–8 (2020). https://doi.org/10.1109/CEC48606.2020.9185901
7. Price, K.V., Storn, R.M., Lampinen, J.A.: Differential Evolution. NCS, Springer, Heidelberg (2005). https://doi.org/10.1007/3-540-31306-0
8. Sallam, K.M., Elsayed, S., Chakrabortty, R.K., Ryan, M.: Improved Multioperator differential evolution algorithm for solving unconstrained problems. In: Proceedings of the IEEE Congress on Evolutionary Computation, CEC, pp. 1–8 (2020). https://doi.org/10.1109/CEC48606.2020.9185577
9. Mohamed, A.W., Hadi, A.A., Mohamed, A.K., Agrawal, P., Kumar, A., Suganthan, P.N.: Problem definitions and evaluation criteria for the CEC 2021 special session and competition on single objective bound constrained numerical optimization. Technical report, Nanyang Technological University, Singapore (2020)

10. Storn, R., Price, K.: Differential evolution - a simple and efficient heuristic for global optimization over continuous spaces. J. Global Optimization **11**, 341–359 (1997). https://doi.org/10.1023/A:1008202821328

11. Tanabe, R., Fukunaga, A.S.: Success-history based parameter adaptation for differential evolution. In: Proceedings of the IEEE Congress on Evolutionary Computation, 71–78. IEEE Press (2013). https://doi.org/10.1109/CEC.2013.6557555

12. Tanabe, R., Fukunaga, A.S.: Improving the search performance of SHADE using linear population size reduction. In: Proceedings of the IEEE Congress on Evolutionary Computation, CEC, Beijing, China, pp. 1658–1665 (2014). https://doi.org/10.1109/CEC.2014.6900380

13. Zhang, J., Sanderson, A.C.: JADE: self-adaptive differential evolution with fast and reliable convergence performance. In: 2007 IEEE Congress on Evolutionary Computation, pp. 2251–2258 2007. https://doi.org/10.1109/CEC.2007.4424751

Optimization and Machine Learning

A New Learnheuristic: Binary SARSA - Sine Cosine Algorithm (BS-SCA)

Marcelo Becerra-Rozas[1]($\boxtimes$)(iD), José Lemus-Romani[2](iD), Broderick Crawford[1](iD), Ricardo Soto[1](iD), Felipe Cisternas-Caneo[1](iD), Andrés Trujillo Embry[1](iD), Máximo Arnao Molina[1](iD), Diego Tapia[1](iD), Mauricio Castillo[1](iD), and José-Miguel Rubio[3]

[1] Pontificia Universidad Católica de Valparaíso, Valparaíso, Chile
{broderick.crawford,ricardo.soto}@pucv.cl,
{marcelo.becerra.r,felipe.cisternas.c,andres.trujillo.e,maximo.arnao.m,
diego.tapia.r,mauricio.castillo.d}@mail.pucv.cl
[2] Pontificia Universidad Católica de Chile, Santiago, Chile
jose.lemus@uc.cl
[3] Universidad Bernardo O'Higgins, Santiago, Chile
josemiguel.rubio@ubo.cl

Abstract. This paper proposes a novel learnheuristic called Binary SARSA - Sine Cosine Algorithm (BS-SCA) for solving combinatorial problems. The BS-SCA is a binary version of Sine Cosine Algorithm (SCA) using SARSA to select a binarization operator. This operator is required due SCA was created to work in continuous domains. The performance of BS-SCA is benchmarked with a Q-learning version of the learnheuristic. The problem tested was the Set Covering Problem and the results show the superiority of our proposal.

Keywords: Learnheuristic · SARSA · Sine Cosine Algorithm · Combinatorial problem

1 Introduction

Optimization problems have been growing in a big way in the last decades, causing the emergence of more metaheuristics (MH) that try to solve NP-Hard combinatorial optimization problems. The premise of the No Free Lunch Theorem [1,2] incentives us to develop increasingly robust optimization algorithms that present and high feasible, quality solutions in reasonable computational times.

To develop more robust algorithms, different techniques used in MH can be distinguished. First, there is the hybridization of mathematical programming with MH or also known as "Matheuristics" [3] There are methods that interrelate MH with simulation problem, also known as "Simheuristics" [4]. There are also hybridization methods between MH techniques that combine their exploration-exploitation components [5]. Currently, the area that is in constant development

© Springer Nature Switzerland AG 2022
B. Dorronsoro et al. (Eds.): META 2021, CCIS 1541, pp. 127–136, 2022.
https://doi.org/10.1007/978-3-030-94216-8_10

and will continue to develop for a couple of years more, learnheuristic is the interaction of MH with learning techniques, where it has been observed in several studies that these techniques support operators in various ways to improve their performance [6–8].

This study presents the incorporation of SARSA to identify a specific action: the selection of binarization strategies to solve binary domain problems, this incorporation has already been carried out, but previously with the whale meta-heuristic [9]. A comparison is made between the proposed implementation of the Binary SARSA-Sine Cosine Algorithm (BS-SCA) and the work presented by Cisternas-Caneo et al. in [10], however, this time the number of evaluated instances is increased. The problem to be solved is the Set Coverage Problem, and after evaluating 45 instances, it can be established that the proposed implementation with SARSA performs statistically significantly better.

The paper is organized as follows: In Sect. 2 we raise points in favor of why it is worthwhile to implement reinforcement learning techniques with swarm intelligence algorithms. In Sect. 3 we present the reinforcement learning techniques belonging to the machine learning area: Q-Learning and SARSA. Our proposed BS-SCA as a new algorithm is presented in Sect. 4. Finally, a proper analysis and discussions are illustrated in Sect. 5, followed by our conclusions and future lines of work in Sect. 6.

2   Swarm-Intelligence Algorithms

Swarm Intelligence Algorithms are regularly based on interesting behaviors found in nature. In particular, in those situations that involve behaviors carried out collectively by some biological systems, such as animals or insects. This is why these algorithms are founded on the study of self-organized and distributed systems, because they manipulate a population of agents with limited individual ability, each of which reacts to its environment and can modify it in order to perform intelligent collective behavior. This ability allows communication between agents, and when they perceive changes in their environment, they interact locally with other agents. In turn, this results in the formation of a global behavior, which allows agents to deal with complicated situations effectively.

MH have different elements depending on the metaphor they represent. Although they are generally composed of an instance, parameters, operators, population, local search, evaluation, initialization, and decision variables [7]. For the definition of the parameters a considerable number of experiments are carried out that will allow their values to be adjusted. This requires the dedication of considerable time and an imbalance between the exploration and exploitation of MH. That is why the need arises for the integration of dynamic elements in the algorithms so that these are adjusted during the execution of the iterations. In this paper, SARSA and Q-Learning are used to perform a dynamic selection of operators.

2.1 Hybrid-Metaheuristics

A hybrid MH is described as the combination of a metaheuristic algorithm and a different learning algorithm, for instance, matheuristics, Machine Learning Programming, Reinforcement Learning (RL) techniques [5]. For this work we will focus on the hybrids generated with RL, where we find two groups: RL supporting MH, or MH supporting RL.

Focusing on the first group mentioned above, two lines of research are shown in the work of García et al. [11]. First, we find the integration of RL techniques as the replacement of an operator, such as the handling of a population, local search, and parameter tuning. Second, is to use RL as a selector of a set of MH, choosing the most appropriate one depending on the problem to be approached.

When using RL as a selector, we can divide this category into three groups. The first is algorithm selection that chooses from a set of techniques for the problem, in order to obtain better performance for a set of similar instances [12]. Secondly we find the hyperheuristic strategies, where their goal is to use the MH to cover a set of problems. And finally we find cooperative strategies, which combine algorithms sequentially with the objective of improving the robustness of the solution.

3 Reinforcement-Learning Techniques

The main objective of these techniques is that the agent manages to learn a policy that is able to maximize the long-term rewards by interacting in turn, with the obtained environment and based on its own experience. The information of the value function will indicate how fruitful is the consequence of the action performed from a state, in other words, how good is the reward. The expected reward function R_t is composed of both the current rewards obtained and the discounted future rewards. The future reward for the passage of time t is given by the following Eq. (1).

$$R_t = \sum_{j=0}^{n} \gamma^j \cdot r_{t+j+1} \qquad (1)$$

Now, based on the above (the search for a policy that maximizes the long-term reward), we in this paper have implemented two recognized techniques in the area of reinforcement learning in order to compare the optimal state-action value function obtained. On the one hand we have: Q-Learning, as used in the work of Cisternas-Caneo et al. [10,13] and SARSA as used in Becerra-Rozas et al. [9].

Q-learning is one of the best known algorithms in the area of reinforcement learning [14] and is an off-policy method, meaning that the agent is independent of the environment where it is executing and chooses the action a that it considers to give it the most value. When the agent selects an action and executes it in the environment a perturbation is generated. The impact of this perturbation is judged through the reward or punishment (r) to decide which is the next state s_{t+1} of the environment. The way to represent the update equation mathematically is Eq. (2):

$$Q_{new}(s_t, a_t) = (1 - \alpha) \cdot Q_{old}(s_t, a_t) + \alpha \cdot [r_n + \gamma \cdot maxQ(s_{t+1}, a_{t+1})] \qquad (2)$$

Where $Q_{new}(s_t, a_t)$ is nominating the reward of the action taken in state s_t and r_n is the reward received when action a_t is taken, $maxQ(s_{t+1}, a_{t+1})$ is the maximum value of the action for the next state, the value of α must be $0 < \alpha \leq 1$ and corresponds to the learning factor. On the other hand, the value of γ must be $0 \leq \gamma \leq 1$ and corresponds to the discount factor. If γ reaches the value of 0, only the immediate reward will be considered, while as it approaches the value 1 the future reward receives greater emphasis relative to the immediate reward.

In contrast, unlike Q-Learning, SARSA [15] is an on-policy algorithm, this tells us that this time, the agent if will be dependent on the execution environment and the next action will be taken based on the value of the current state-action. For this reason, it is often said that SARSA is a more conservative algorithm than Q-Learning and this, in turn, allows it to learn faster. Based on this, the state-action value update equation is defined as in Eq. (3):

$$Q(s_t, a_t) \longleftarrow Q(s_t, a_t) + \alpha \cdot [r + \gamma \cdot Q(s_{t+1}, a_{t+1}) - Q(s_t, a_t)] \qquad (3)$$

3.1   Reward Function

A good balance of reward and punishment results in an equal variety in action selection, which makes the optimal action identified more trustworthy. For this reason, we will use a simplified version of Xu and Pi's work [16]. The actions of our smart selector will be rewarded based on this version, which considers as reward value +1 when fitness is improved or 0 otherwise. In Eq. (4) contained in Table (1) we can see the above (Table 1).

Table 1. Types of rewards

Reference	Reward function	
[16]	$r_n = \begin{cases} +1, & \textit{If the current action improves fitness} \\ 0, & \textit{otherwise.} \end{cases}$	(4)

4   Binary SARSA - Sine Cosine Algorithm

Sine Cosine Algorithm (SCA) [17] is a swarm metaheuristic of recent interest to researchers for solving complex optimisation problems. While it is a metaheuristic that provides great results, it still falls into the classic problem of swarm metaheuristics, falling into premature convergences which implies falling into local optima [18]. Recent works [10,13,19], the authors propose ambidextrous metaheuristics [20,21] where their main objective is to improve decision making during the optimisation process, which translates into improving the exploration and exploitation balance, i.e. avoiding premature convergence and thus improving the solutions obtained.

The authors in [10,19] propose the incorporation of Q-Learning to Sine Cosine Algorithm as an intelligent binarization schemes selector mechanism to solve discrete optimisation problems [22]. Our proposal is to replace Q-Learning by another intelligent selector mechanism such as SARSA and compare both techniques performing the same task, selecting binarisation schemes with the aim of improving the major problem of SCA, premature convergence.

As proposed in [19], the states used in SARSA are the phases of the MH, i.e., exploration and exploitation. The estimation of these states is done by means of diversity metrics which allow quantifying the dispersion of individuals in the search space. The metric used in this work is the Dimensional-Hussain Diversity [23] and is defined as follows:

$$Div = \frac{1}{l \cdot n} \sum_{d=1}^{l} \sum_{i=1}^{n} |\bar{x}^d - x_i^d| \tag{5}$$

Where n is the number of search agents in the population X, $\bar{x}^d$ is average of the d-th dimension, and l is the number of dimension of the optimization problem.

This diversity quantification is calculated iteration by iteration and to determine whether the population has an exploration or exploitation behavior the equations proposed by Morales-Castañeda et al. in [24] are used. There they propose that the percentage of exploration (XPL%) and the percentage of exploitation (XPT%) is given as follows:

$$XPL\% = \left(\frac{Div_t}{Div_{max}} \right) \times 100 \quad , \quad XPT\% = \left(\frac{|Div_t - Div_{max}|}{Div_{max}} \right) \times 100 \tag{6}$$

By obtaining these percentages, the phase in which the MH is found is determined as follows:

$$next\ state = \begin{cases} Exploration\ if\ XPL\% \geq XPT\% \\ Exploitation\ if\ XPL\% < XPT\% \end{cases} \tag{7}$$

The proposal of this work is shown in Algorithm 1. In line 1 we initialise the Q-values of the Q-Table, in lines 4–5 we determine the initial state (exploration or exploitation) of SARSA, in line 7 we select an action from the Q-Table for the corresponding state, in line 16 we execute the selected action and observe its consequences from the obtained fitness, in lines 17–18 we determine the next state of SARSA, and finally in line 19 we update the Q-value of the selected action from the SARSA Eq. (3).

Algorithm 1. Binary S-Sine Cosine Algorithm

Input: The population $X = \{X_1, X_2, ..., X_n\}$
Output: The updated population $X' = \{X'_1, X'_2, ..., X'_n\}$ and X_{best}
1: **Initialize Q-Table with** q_0
2: Initialize random population X
3: Set initial r_1
4: **Calculate Initial Population Diversity (X) using equation (5)**
5: **Define the initial state using equation (7)**
6: **for** *iteration* (t) **do**
7:     a : **Select action from Q-Table**
8:     **for** *solution* (i) **do**
9:         Evaluate solution X_i in the objective function
10:         **for** *dimension* (j) **do**
11:             Update P_j^t, where $P_j^t = X_{best,j}$
12:             Randomly generate the value of r_2, r_3, r_4
13:             Update the position of $X_{i,j}$
14:         **end for**
15:     **end for**
16:     **Binarization X with action a and apply reward function**
17:     **Calculate Population Diversity (X) using equation (5)**
18:     **Define the next state using equation (7)**
19:     **Update Q-Table using SARSA equation (3)**
20:     Update r_1
21:     Update X_{best}
22: **end for**
23: Return the updated population X where X_{best} is the best result

---

5   Experimental Results

We present the results in the table (2). The table is composed such that: in the first column is the name of the instance, in the second, the known optimal value of each instance. For the remaining columns, the three subsequent columns indicate: the best result achieved in each instance (Best), the average of the results (Avg) and the relative percentage deviation (RPD) according to the Eq. (8). These three columns are replicated for both implemented versions. The comparison is made with the re-implementation of the algorithm proposed by Cisternas-Caneo et al. [10], however, this time with more instances. The sum of all columns and the p-value of the Wilcoxon Mann-Whitney test [25] are presented in the last two rows. To establish which of the two hybridized versions is superior, the test allows us to evaluate whether the results obtained differ considerably (Table 2).

$$\text{RPD} = \frac{100 \cdot (Best - Opt)}{Opt}. \tag{8}$$

The total number of instances used to solve the Set Covering Problem with Beasley's OR-Library instances was 45. These cases were run with 40 populations and 1000 iterations, with a total of 40,000 calls to the objective function, as used

Table 2. Results obtained solving SCP by BS-SCA and BQ-SCA

Inst.	Opt.	BS-SCA			BQ-SCA		
		Best	Avg	RPD	Best	Avg	RPD
4.1	429	**432**	434.0	0.7	435	442.72	1.4
4.2	512	**527**	535.9	2.93	537	553.71	4.88
4.3	516	**524**	529.0	1.55	534	552.03	3.49
4.4	494	**502**	515.12	1.62	514	530.44	4.05
4.5	512	**524**	530.14	2.34	537	553.17	4.88
4.6	560	**564**	570.56	0.71	573	588.68	2.32
4.7	430	**435**	439.25	1.16	441	449.77	2.56
4.8	492	**500**	502.57	1.63	509	516.39	3.46
4.9	641	**665**	677.14	3.74	683	697.48	6.55
4.10	514	**518**	519.57	0.78	521	533.88	1.36
5.1	253	**256**	266.22	1.19	264	272.75	4.35
5.2	302	**318**	326.27	5.3	327	335.58	8.28
5.3	226	**230**	231.1	1.77	**230**	235.62	1.77
5.4	242	**247**	250.22	2.07	250	254.6	3.31
5.5	211	**213**	215.62	0.95	218	221.46	3.32
5.6	213	**218**	222.12	2.35	221	231.26	3.76
5.7	293	**297**	305.11	1.37	304	316.4	3.75
5.8	288	**290**	294.56	0.69	296	301.32	2.78
5.9	279	**283**	285.14	1.43	284	293.42	1.79
5.10	265	**271**	273.0	2.26	274	281.35	3.4
6.1	138	**143**	146.0	3.62	144	148.16	4.35
6.2	146	**151**	152.56	3.42	152	159.06	4.11
6.3	145	**148**	149.5	2.07	149	151.29	2.76
6.4	131	**131**	133.6	0.0	133	136.03	1.53
6.5	161	**165**	171.82	2.48	173	183.26	7.45
a.1	253	**260**	264.22	2.77	266	269.42	5.14
a.2	252	**254**	266.2	0.79	267	273.8	5.95
a.3	232	**238**	244.25	2.59	245	248.87	5.6
a.4	234	**241**	246.5	2.99	245	252.61	4.7
a.5	236	**242**	245.0	2.54	247	251.27	4.66
b.1	69	**69**	70.9	0.0	71	72.68	2.9
b.2	76	**76**	77.8	0.0	78	81.35	2.63
b.3	80	**80**	84.4	0.0	82	83.87	2.5
b.4	79	**82**	83.3	3.8	83	84.9	5.06
b.5	72	**72**	73.12	0.0	73	75.03	1.39
c.1	227	**237**	240.5	4.41	246	251.85	8.37
c.2	219	**230**	235.44	5.02	237	242.89	8.22
c.3	243	**252**	254.88	3.7	259	263.25	6.58
c.4	219	**228**	232.25	4.11	230	236.1	5.02
c.5	215	**222**	225.5	3.26	229	234.2	6.51
d.1	60	**62**	64.6	3.33	64	65.97	6.67
d.2	66	**67**	73.27	1.52	69	69.97	4.55
d.3	72	**75**	81.8	4.17	76	78.86	5.56
d.4	62	**62**	64.6.	0.0	63	64.16	1.61
d.5	61	**62**	69.89	1.64	64	66.35	4.92
Average		**259.18**	263.88	2.11	264.38	271.27	4.23
p-value							0.00

in [26]. The code was written in Python 3.8 and executed using the free Google Colaboraty service [27]. The following parameters were specified for the SARSA and Q-Learning algorithms: $\gamma = 0.4$ and $\alpha = 0.1$.

The exploration-exploitation graphs obtained: Fig. 1 and 2 according to Sect. 4, do not show similar behaviors to those presented by Morales-Castañeda et al. in [24], despite the fact that our proposal does not have similarities to the graphs presented by them, when observing the results obtained it is determined that they are not random algorithms since they present variations in their exploration percentages.

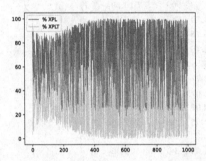

Fig. 1. SCP - Exploration and Exploitation Graphic of instance 4.7 version BS-SCA

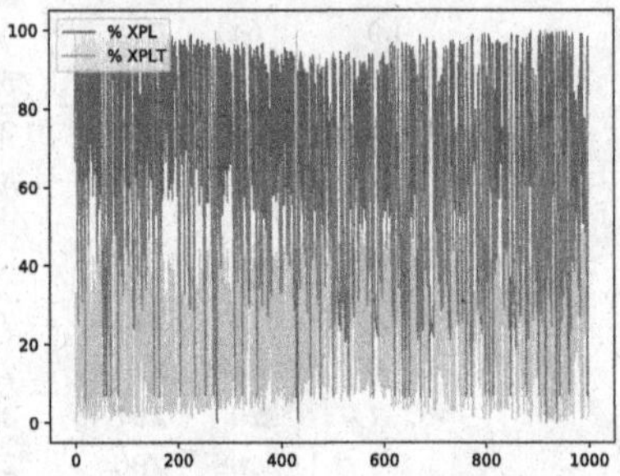

Fig. 2. SCP - Exploration and Exploitation Graphic of instance 4.7 version BQ-SCA

6 Conclusion

The results are encouraging because the performance of a binarization selector reduces tuning times by not having to evaluate combinations of various binarization schemes in the literature.

SARSA has proven to be efficient as an intelligent selector of binarization techniques when evaluating it in the sine-cosine algorithm with the 45 instances present in OR-library of the Set Covering Problem. This can be noticed when comparing it with the version applied by Q-Learning, since we obtain better results in 44 of the 45. On the other hand, these results are statistically significantly better when performing the Wilcoxon-Mann-Whitney test.

We can also notice that when observing in detail the graphs used for exploration and exploitation there is a similar convergence, but with BS-SCA the values tend to have smaller magnitude changes, with higher occurrence, and much more defined which may indicate that they are more efficient in solving this problem.

As future work, in addition to the implementation of SARSA in other metaheuristic techniques, we will also seek to implement other reinforced learning techniques, other transfer functions and ways to binarize so that our intelligent selector has more to choose from. It is also necessary to parameterize the results obtained through other types of graphs or metrics. Although it provides valuable and useful information about the search process, it is still necessary a comparative metric and in turn, that this same metric can be incorporated in the learning process of the agent.

Acknowledgement. Broderick Crawford is supported by Grant CONICYT/ FONDECYT/REGULAR/1210810. Broderick Crawford, Ricardo Soto and Marcelo Becerra-Rozas are supported by Grant Nucleo de Investigacion en Data Analytics/VRIEA/PUCV/039.432/2020. Ricardo Soto is supported by Grant CONICYT/FONDECYT/REGULAR/1190129. Marcelo Becerra-Rozas are supported by Grant DI Investigación Interdisciplinaria del Pregrado/VRIEA/PUCV/039.421/2021. Marcelo Becerra-Rozas is supported by National Agency for Research and Development (ANID)/Scholarship Program/DOCTORADO NACIONAL/2021-21210740. José Lemus-Romani is supported by National Agency for Research and Development (ANID)/Scholarship Program/DOCTORADO NACIONAL/2019-21191692.

References

1. Wolpert, D.H., Macready, W.G.: No free lunch theorems for optimization. IEEE Trans. Evol. Comput. **1**, 67–82 (1997)
2. Wolpert, D.H., Macready, W.G.: Coevolutionary free lunches. IEEE Trans. Evol. Comput. **9**, 721–735 (2005)
3. Voss, S., Maniezzo, V., Stützle, T.: Matheuristics: Hybridizing Metaheuristics and Mathematical Programming (Annals of Information Systems). Springer, Boston (2009). https://doi.org/10.1007/978-1-4419-1306-7
4. Juan, A.A., Faulin, J., Grasman, S.E., Rabe, M., Figueira, G.: A review of simheuristics: extending metaheuristics to deal with stochastic combinatorial optimization problems. Oper. Res. Perspect. **2**, 62–72 (2015)
5. Talbi, E.-G.: Combining metaheuristics with mathematical programming, constraint programming and machine learning. Ann. Oper. Res. **240**(1), 171–215 (2015). https://doi.org/10.1007/s10479-015-2034-y
6. Talbi, E.G.: Machine learning into metaheuristics: a survey and taxonomy of data-driven metaheuristics. (2020)
7. Song, H., Triguero, I., Özcan, E.: A review on the self and dual interactions between machine learning and optimisation. Prog. Artif. Intell. **8**(2), 143–165 (2019). https://doi.org/10.1007/s13748-019-00185-z
8. Calvet, L., de Armas, J., Masip, D., Juan, A.A.: Learnheuristics: hybridizing metaheuristics with machine learning for optimization with dynamic inputs. Open Math. **15**, 261–280 (2017)
9. Becerra-Rozas, M., et al.: Reinforcement learning based whale optimizer. In: Gervasi, O., et al. (eds.) ICCSA 2021. LNCS, vol. 12957, pp. 205–219. Springer, Cham (2021). https://doi.org/10.1007/978-3-030-87013-3_16
10. Cisternas-Caneo, F., et al.: A data-driven dynamic discretization framework to solve combinatorial problems using continuous metaheuristics. In: Abraham, A., Sasaki, H., Rios, R., Gandhi, N., Singh, U., Ma, K. (eds.) IBICA 2020. AISC, vol. 1372, pp. 76–85. Springer, Cham (2021). https://doi.org/10.1007/978-3-030-73603-3_7
11. García, J., et al.: A db-scan binarization algorithm applied to matrix covering problems. Comput. Intell. Neurosci. **2019**, 1–16 (2019)

12. de León, A.D., Lalla-Ruiz, E., Melián-Batista, B., Moreno-Vega, J.M.: A machine learning-based system for berth scheduling at bulk terminals. Expert Syst. Appl. **87**, 170–182 (2017)

13. Tapia, D., et al.: Embedding q-learning in the selection of metaheuristic operators: the enhanced binary grey wolf optimizer case. In: 2021 IEEE International Conference on Automation/XXIV Congress of the Chilean Association of Automatic Control (ICA-ACCA), pp. 1–6 (2021)

14. Watkins, C.J., Dayan, P.: Q-learning. Mach. Learn. **8**, 279–292 (1992)

15. Sutton, R.S., Barto, A.G.: Reinforcement Learning: An Introduction. MIT Press (2018)

16. Xu, Y., Pi, D.: A reinforcement learning-based communication topology in particle swarm optimization. Neural Comput. Appl. **32**(14), 10007–10032 (2019). https://doi.org/10.1007/s00521-019-04527-9

17. Mirjalili, S.: SCA: a sine cosine algorithm for solving optimization problems. Knowl.-Based Syst. **96**, 120–133 (2016)

18. Kai Feng, Z.: A modified sine cosine algorithm for accurate global optimization of numerical functions and multiple hydropower reservoirs operation. Knowl. Based Syst. **208**, 106461 (2020)

19. Crawford, B., et al.: A comparison of learnheuristics using different reward functions to solve the set covering problem. In: Dorronsoro, B., Amodeo, L., Pavone, M., Ruiz, P. (eds.) OLA 2021. CCIS, vol. 1443, pp. 74–85. Springer, Cham (2021). https://doi.org/10.1007/978-3-030-85672-4_6

20. Crawford, B., León de la Barra, C.: Los algoritmos ambidiestros (2020). https://www.mercuriovalpo.cl/impresa/2020/07/13/full/cuerpo-principal/15/. Accedia 12 Feb 2021

21. Lemus-Romani, J., et al.: Ambidextrous socio-cultural algorithms. In: Gervasi, O., et al. (eds.) ICCSA 2020. LNCS, vol. 12254, pp. 923–938. Springer, Cham (2020). https://doi.org/10.1007/978-3-030-58817-5_65

22. Crawford, B., Soto, R., Astorga, G., García, J., Castro, C., Paredes, F.: Putting continuous metaheuristics to work in binary search spaces. Complexity **2017**, 1–19 (2017)

23. Hussain, K., Zhu, W., Salleh, M.N.M.: Long-term memory Harris' hawk optimization for high dimensional and optimal power flow problems. IEEE Access **7**, 147596–147616 (2019)

24. Morales-Castañeda, B., Zaldivar, D., Cuevas, E., Fausto, F., Rodríguez, A.: A better balance in metaheuristic algorithms: does it exist? Swarm Evol. Comput. **54**, 100671 (2020)

25. Mann, H.B., Whitney, D.R.: On a test of whether one of two random variables is stochastically larger than the other. Ann. Math. Stat. **18**, 50–60 (1947)

26. Lanza-Gutierrez, J.M., Crawford, B., Soto, R., Berrios, N., Gomez-Pulido, J.A., Paredes, F.: Analyzing the effects of binarization techniques when solving the set covering problem through swarm optimization. Expert Syst. Appl. **70**, 67–82 (2017)

27. Bisong, E.: Google colaboratory. In: Building Machine Learning and Deep Learning Models on Google Cloud Platform, pp. 59–64. Apress, Berkeley, CA (2019). https://doi.org/10.1007/978-1-4842-4470-8_7

Minimum Rule-Repair Algorithm for Supervised Learning Classifier Systems on Real-Valued Classification Tasks

Koki Hamasaki[✉] and Masaya Nakata[✉]

Department of Electrical Engineering and Computer Science,
Yokohama National University, Yokohama, Kanagawa, Japan
{hamasaki-koki-pt,nakata-masaya-tb}@ynu.ac.jp

Abstract. This paper proposes a simple rule-repair algorithm for the UCS classifier system with real-valued inputs on classification problems. Our concept is to repair inaccurate rules with a possible minimum reduction of the rule-generality in order to avoid the problematic cover-delete cycle. We identify the following two principles to achieve this purpose; 1) to repair the rule-condition to omit one incorrect input from a matching space represented by its rule-condition; 2) to repair either a lower value or an upper value for one dimension x_i. Experiments confirmed the adequacy of those principles. Consequently, UCS with our rule-repair algorithm successfully boosts the performance while preventing the increase of the population size.

Keywords: Machine learning · Learning Classifier System · Supervised learning

1 Introduction

Learning Classifier Systems [1] (LCSs) are a paradigm of evolutionary rule-based learning methods. LCSs intend to produce accurate, maximally general, and thus explainable rules [2,3]. Relying on this advantage, many works have applied LCSs to data-mining tasks [4,5]. Technically, LCSs are designed to generate a minimal rule-set that determines plausible outputs for given inputs. Thus, each rule should accurately predict an output while covering as many inputs as possible. For this purpose, rule-based learning evaluates a rule-fitness through interaction with an environment, and then evolutionary computation, e.g., GA, generatively refines rules with fitness guidance.

While many branches of LCSs have been proposed thus far [6,7], most works have extended either one of the two basic LCSs: the XCS classifier system [8] and the UCS classifier system [9]. XCS is based on a reinforcement learning (RL) approach, and thus it is suitable for RL problem domains, e.g., online-control [10]. XCS evaluates the rule-fitness with reward signals. In contrast, UCS is an extension of XCS, and it uses a supervised learning approach, where both

© Springer Nature Switzerland AG 2022
B. Dorronsoro et al. (Eds.): META 2021, CCIS 1541, pp. 137–151, 2022.
https://doi.org/10.1007/978-3-030-94216-8_11

an input and the correct output for it are sent to the system. Thus, UCS can be suitable for supervised learning tasks, e.g., classification [9,11]. Note that XCS and UCS commonly use the steady-state GA as a rule-evolution scheme. Consider the LCS's advantage aforementioned, this paper studies UCS as a data-mining tool for classification.

However, a restriction of LCSs (including UCS) is in less scalability of the system performance against the input space size. For instance, the LCS performance significantly degrades when dealing with high-dimensional and/or real-valued inputs [12–14]. To tackle this issue, Debie provided a theoretical insight for UCS on high-dimensional problems [15]. He also proposed an ensemble learning scheme of UCS to boost the performance on real-valued classification tasks [16]. Urabanowicz introduced some heuristics for UCS to improve the efficiency of rule-evolution (i.e., ExSTraCS) [17]. ExSTraCS successfully solves the 135-bit multiplexer problem with binary inputs. Some modern works revealed the impacts of the lexicase selection [18] and the fine-tuning for hyper-parameters [19] on the UCS framework. In addition, dimensional reduction techniques were used in XCS, e.g., feature selection [20,21] and deep auto-encoder approaches [13,14]; those approaches can be extended to the UCS framework.

Although various extensions of UCS have been considered as aforementioned, there are very few works that intend to repair inaccurate rules for the real-valued UCS framework. A possible reason is that UCS is originally designed to evolve only accurate rules to construct a *best action map* [9]. However, as another critical reason common to XCS, UCS should be designed to maintain a low frequency of the rule-production to avoid a problematic cover-delete cycle [22,23]. For instance, the steady-state GA produces only two offspring rules per generation, which results in a fundamental inefficiency of UCS. Here, the cover-delete cycle is one of the major difficulties to design online learning-based LCSs, meaning that insufficiently-trained rules may be deleted due to a high frequency of rule-production; and this cycle may frequently occur when each rule is less general under a limited population size. Thus, a rule-repair strategy can be a possible reason to provoke the cover-delete cycle.

Note that there are some rule-repair algorithms for XCS. In [24], Lanzi proposed a *specify* operator for XCS with binary-input problems. His concept is to repair inaccurate rules identified by the XCS's reinforcement learning scheme. In detail, the specify operator replaces some don't care bits involved in a rule-condition with specific values of the input. This operator was applied to the UCS framework [9]. In [25], Iqbal presented a GP-based XCS and its rule-repair algorithm for GP-based rule expression. Tadokoro introduced a local covering operator [26]. While this operator does not intend to repair inaccurate rules, it produces new initial rules with a similar concept to the specify operator. Although those related works show the effectiveness of rule-repair algorithms, they have not been designed for UCS with real-valued inputs.

Accordingly, this paper presents a minimum rule-repair algorithm for UCS with real-valued inputs on classification problems. Our rule-repair algorithm intends to improve the performance by 1) boosting the classification accuracy (i.e., the rule-fitness) of inaccurate rules and by 2) increasing the frequency of rule-reproductions. Besides, we design our algorithm based on a minimum rule-repair concept to avoid the problematic cover-delete cycle. That is, our algorithm repairs the rule-condition with the minimum reduction of its rule-generality; the rule-condition is repaired so that it excludes one incorrect input from a subspace covered by its rule-condition.

This paper is organized as follows. Section 2 describes the UCS framework for real-valued inputs. Section 3 introduces our rule-repair algorithm. Section 4 tests UCS with our rule-repair algorithm on real-valued benchmark classification problems. Section 5 empirically validates our hypothetical insights. Finally, in Sect. 6, we summarize our contributions with future directions.

2   UCS for Real-Valued Inputs

This section gives a description of the UCS framework for real-valued inputs $x = [x_1, x_2, \cdots, x_d]$, where d is the problem dimension and $x_i \in [0,1] (i = \{1, 2, \cdots, d\})$. This paper employs a lower-upper representation as a rule-condition for real-valued inputs [27]. The rule-condition with this coding represents a d-dimension hyperrectangle as its matching sub-space on the input space; and thus its rule-generality can be measured with a volume of its hyper-rectangle. Note that this paper denotes a uniformly-sampled random value as r, and $r \in [0,1]$, if not stated differently; and all rs used in equations are independently sampled. Note also that we introduce new mathematical notations for the UCS framework, which is exactly the same as in the original working of UCS [9].

2.1   Rule Parameters

A rule cl consists of a condition $C = \{c_1, c_2, \cdots c_d\}$ and an action A, where a sub-condition c_i involves a lower l_i and an upper u_i both used for x_i, i.e., $c_i = [l_i, u_i]$ ($l_i \le u_i$, $l_i, u_i \in [0,1]$). A rule cl can be matched to x if and only if $l_i \le x_i \le u_i, \forall i \in \{1, 2, \cdots, d\}$, simply denoted by $x \in C$ in this paper. The action A represents a class when its rule is executed.

The rule cl also has the following five main parameters; the number of correct classification $ct \in \mathbb{N}_0$, which represents how many times cl belongs to $[C]$; the accuracy $acc \in [0,1]$, which is a classification accuracy of cl; the rule-fitness $F \in [0,1]$, which is calculated from acc; the experience $exp \in \mathbb{N}_0$, which represents the number of parameter-update times; the numerosity $num \in \mathbb{N}_0$; which is the number of subsumed rules to cl by a subsumption operator (see Sect. 2.2).

Suppose two rules cl_1 and cl_2 both having the same action, cl_1 can be more general than cl_2 if and only if $l_{1,i} \le l_{2,i} \wedge u_{2,i} \le u_{1,i}, \forall i \in \{1, 2, \cdots, d\}$, simply denoted by $cl_2.C \subset cl_1.C$ in this paper. All rules are contained in a population $[P]$ with the maximum population size N.

2.2   Framework

The UCS framework is composed of the training phase and the test phase. During training, UCS activates rule-parameter updates and the steady-state GA in order to produce the optimal rule-set as a solution. During the test phase, it only determines an output based on the trained rule-set.

2.2.1   Training Phase

At the initial iteration $t = 0$, UCS builds the population $[P]$ as an empty set. For $t \leftarrow t+1$, UCS receives an input x together with its correct class A^*. Then, it builds a match set $[M]$ consisting rules matched to x, given by;

$$[M] = \{cl \in [P] \mid x \in cl.C\}. \tag{1}$$

Then, UCS further builds a correct set $[C]$ and an incorrect set $[!C]$, given by;

$$\begin{cases} [C] & = \{cl \in [M] \mid cl.A = A^*\}, \\ [!C] & = \{cl \in [M] \mid cl.A \neq A^*\}. \end{cases} \tag{2}$$

Thus, $[C]$ and $[!C]$ are composed of (temporarily) accurate rules and inaccurate rules, respectively. If $[C]$ is empty, the covering operator takes place to produce a new rule with an initial setting $\{A = A^*, ct = 0, exp = 0, F = 0.01, num = 1\}$; and l_i and u_i for $c_i \in C$ are initialized as;

$$\begin{cases} l_i & = x_i - r \cdot s_0, \\ u_i & = x_i + r \cdot s_0, \end{cases} \tag{3}$$

where $s_0 \in [0,1]$ is a hyperparameter that controls the initial rule-generality. Thus, $x \in C$ is always satisfied. Note that $0 \leq l_i \leq u_i \leq 1$.

Next, UCS updates rule parameters. First, ct is updated as $ct \leftarrow ct + 1$ for each rule in $[C]$. Next, for all rules in $[M]$, exp, acc, and F are updated. In detail, exp is updated as $exp \leftarrow exp + 1$ to count the update time; then, acc is updated by;

$$acc = \frac{ct}{exp}. \tag{4}$$

Thus, acc represents the classification accuracy of cl. Then, the fitness F is updated with exponential reduction, given by;

$$F = (acc)^\nu, \tag{5}$$

where ν controls a selection bias in the steady-state GA.

Finally, the steady-state GA is applied to $[C]$ to generate plausibly better rules. First, UCS selects two parent rules from $[C]$; and it produces two offspring rules cl_1, cl_2 as copies of the corresponding parent rules except for $\{ct = 0, exp = 0, F = 0.01, num = 1\}$. Then, a crossover operator is activated with a probability χ; and this paper employs the uniform crossover. If the crossover is activated, it may swap $c_{1,i} \in cl_1.C$ for $c_{2,i} \in cl_2.C$ with a probability 0.5 for each $i = \{1, 2, \cdots, d\}$. Next, the mutation operator is also applied to each sub-condition of $cl_*.C$ (i.e., $c_{*,i}$, with a probability μ ($*$ can be 1 and 2)). In detail, $l_{*,i}$ and $\dot{u}_{*,i}$ for $c_{*,i}$ may be mutated as;

$$l_{*,i} \leftarrow \begin{cases} l_{*,i} - r \cdot m_0 & r < 0.5, \\ l_{*,i} + r \cdot m_0 & \text{otherwise,} \end{cases} \tag{6}$$

$$u_{*,i} \leftarrow \begin{cases} u_{*,i} - r \cdot m_0 & r < 0.5, \\ u_{*,i} + r \cdot m_0 & \text{otherwise,} \end{cases} \tag{7}$$

where m_0 is a hyperparameter that controls a degree of the rule-generality; again, $0 \leq l_{*,i} \leq u_{*,i} \leq 1$. Then, two offspring rules are inserted to $[P]$; and rules may be deleted if the population size $[P]$ exceeds N. This paper uses the tournament selection with a tournament size τ.

A subsumption operator may be applied to the rules in $[C]$ after updating the rule parameters or to offspring rules after the steady-state GA. A rule can be subsumed by a more general rule than it, provided that the more general rule is reliably accurate and sufficiently updated (i.e., $acc > acc_0 \land exp > \theta_{sub}$); $acc_0 \in [0, 1]$ defines the minimum classification accuracy the maximally accurate rules must have; and $\theta_{sub} \in \mathbb{N}$ defines the minimum update time. In detain, for each rule cl in $[C]$ except for maximally accurate, maximally general rules cl^*, cl^* subsumes cl if $cl.C \subset cl^*.C \land cl.A = cl^*.A$; then, the numerosity of cl^* is updated as $cl^*.num \leftarrow cl^*.num + cl.num$, and cl is deleted from $[P]$.

2.2.2 Test Phase

For a given input x, UCS builds the match set $[M]$, where all actions a existed in $[M]$ are contained in $[A_M]$. Then, for each action $a \in [A_M]$, it builds a subset $[M_a]$ which consists of rules having the action a, given by $[M_a] = \{cl \in [M] \mid cl.A = a\}$. Finally, it outputs the best action A' having the highest fitness, that is,

$$A' = \arg \max_{a \in [A_M]} \sum_{cl \in [M_a]} cl.F. \tag{8}$$

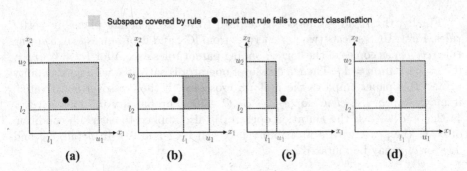

Fig. 1. Examples of possible repair patterns of the rule-condition to eliminate a misclassified input (denoted by the black dot). "A" represents the original subspace covered by a rule; "B", "C", and "D" represent repaired subspaces by repairing $\{l_1, u_2\}$, u_1, and l_1, respectively. "D" can have the hightest rule-generality in those examples.

3   Minimum Rule-Repair Algorithm

In this section, we first introduce our concept of the minimum rule-repair algorithm. Then, the detailed algorithm is described.

3.1   Concept

As described in the previous section, given x at iteration t, the UCS framework does not intend to utilize rules temporarily identified as inaccurate, (i.e., $cl' \in [!C])$. However, such inaccurate rules may contribute to the correct classification for other matched inputs. Consider cl' misclassifies some inputs x', its classification accuracy ($cl'.acc$), certainly improves if $cl'.C$ is repaired to match inputs except for x's. Thus, acc tends to improve by reducing the rule-generality.

However, this strategy (i.e., to reduce the rule-generality), provokes the problematic cover-delete cycle under a restricted population size, as noted in Sect. 1. Thus, we here consider a conservative approach to design our rule-repair algorithm. Our algorithm is designed to repair cl''s rule-condition with a possible minimum reduction of the rule-generality. Specifically, given x at t, we repair $cl'.C$ to eliminate x from its matching sub-space represented by its rule-condition. That is, we target one single input for each repair to avoid a drastic reduction of the rule-generality. In this case, we can still consider various repair patterns of the rule-condition. Figure 1 shows possible examples of repair patterns on two-dimensional inputs $x = [x_1, x_2]$; As shown in this figure, we can suppose possible repair patters ("B", "C", and "D"); however, "D" can have the highest rule-generality in those patterns. Technically, we do not need to repair both l_i and u_i and/or more than one dimension x_i to achieve the minimum reduction of the rule-generality.

Thus, our minimum rule-repair algorithm is designed to satisfy the following two conditions; 1) to repair the rule-condition to eliminate x from its matching sub-space, and 2) to repair either l_i or u_i for one dimension x_i.

3.2 Algorithm

Our rule-repair algorithm is activated after the rule-parameter update in the UCS framework. First, we add a new rule-parameter $rt \in \mathbb{N}_0$, which denotes the latest update time its rule was repaired. The initial value of rt is set to 0 when a rule is generated by the covering operator or the steady-state GA. Then, it selects and then repairs only sufficiently-updated rules. In detail, UCS with our algorithm builds a repair set $[R]$, given by;

$$[R] = \{cl \in [!C] \mid cl.exp - cl.rt > \theta_{sub}\} . \tag{9}$$

Thus, once a rule is repaired, we temporarily remove its rule from candidates for repair until its rule-quality is estimated trustworthy. In other words, this boosts a stable convergence of rule-parameters, and it also prevents a drastic reduction of the rule-generality.

Next, for each $cl \in [R]$, we randomly select a dimension index k from $[1, 2, \cdots, d]$ to decide a target sub-condition c_k for repair; and then it generates new upper/lower candidates $\hat{l}_k, \hat{u}_k$ as;

$$\begin{cases} \hat{l}_k &= x_k + D , \\ \hat{u}_k &= x_k - D , \end{cases} \tag{10}$$

where a hyperparameter $D \in [0,1]$ controls the minimum distance between a specific input value x_k and $\hat{l}_k/\hat{u}_k$. Thus, cl's rule condition can be guaranteed that it does not match x at t when either $l_k = \hat{l}_k$ or $u_k = \hat{u}_k$. Note that D should be set to a relatively small value, e.g., $D \leq 0.01$, to maintain a small reduction of the rule-generality; in Sect. 5, we empirically reveal the impact of D. Next, we further select either $\hat{l}_k$ or $\hat{u}_k$ to maintain the rule-generality as possible. Since the other sub-conditions $c_i(i \neq k)$ are not repaired, a volume of the hyperrectangle specified by $cl.C$ changes dependent on the length of c_k (i.e., $u_k - l_k$). Thus, to maximize the rule-generality, it is sufficient that we can use the candidate maximizing the length of c_k, that is;

$$\begin{cases} l_k \leftarrow \hat{l}_k & \text{if } u_k - \hat{l}_k > \hat{u}_k - l_k, \\ u_k \leftarrow \hat{u}_k & \text{otherwise.} \end{cases} \tag{11}$$

Note that as an exceptional case, we forcedly set u_k to $\hat{u}_k$ if $\hat{l}_k > u_k$, and vice versa; if $\hat{l}_k > u_k$ and $\hat{u}_k < l_k$, we skip to repair the rule-condition, but this is the extremely rare case in our experiments. Finally, rt is updated as $rt \leftarrow exp$. Algorithm 1 shows the pseudo-code of our algorithm.

4 Experiment

This section tests our rule-repair algorithm on two real-valued benchmark classification tasks: the real-valued multiplexer problem (RMUX) and the real-valued majority-on problem (RMOP).

Algorithm 1 Minimum rule-repair algorithm

1: **Input:** x, $[!C]$
2: **for each** $cl \in [!C]$ **do**
3:     **if** $cl.exp - cl.rt > \theta_{sub}$ **then**
4:         $k \leftarrow$ *randomly sampled from* $[1, 2, \cdots, d]$
5:         $\hat{l}_k = x_k + D$
6:         $\hat{u}_k = x_k - D$
7:         **if** $u_k - \hat{l}_k > \hat{u}_k - l_k$ **then**
8:             $l_k \leftarrow \hat{l}_k$
9:         **else**
10:             $u_k \leftarrow \hat{u}_k$
11:         **end if**
12:         $cl.rt \leftarrow cl.exp$
13:     **end if**
14: **end for**

4.1   Benchmark Problems

4.1.1   Real-Valued Multiplexer Problem

The n-bit multiplexer problem (n-MUX) is originally used as a binary input classification problem to validate the generalization capacity of XCS [8]. It has been extended for real-valued classification tasks [12,28]. The n-MUX is defined over a binary string of $n = k + 2^k$; a decimal number of the first k bits represents a position of one of the remaining 2^k bits. Then, the correct class is the bit pointed to by the first k bits. In n-RMUX, each attribute x_i is binarized at the common boundary 0.5 (i.e., 0 if $x_i < 0.5$ or 1 if $x_i \geq 0.5$) and then a correct class of the real-valued input is determined with the same procedure of the n-MUX. Note that, the binarization boundary is not a critical factor dependent on the problem difficulty, as the steady-state GA is designed to independently change the sub-condition c_i based on only its corresponding attribute. This paper uses $\{6, 11, 20, 37\}$-RMUXs to evaluate the scalability of the algorithm.

4.1.2   Real-Valued Majority-On Problem

The n-bit majority-on problem (n-MOP) is also originally defined with binary inputs [25]. In n-MOP, if the number of "1" exceeds the number of "0", the correct class is 1; otherwise 0. This paper extends n-MOP to a real-valued classification task; each attribute x_i is also binarized at the common boundary 0.5. The n-MOP and n-RMOP are highly overlapping problems, where LCSs often suffer to improve the performance [25]. This paper uses $\{11, 15\}$-RMOPs.

4.2   Experimental Settings

We employ the following experimental paradigm. One iteration involves a set of one training input and one test input of the problem. For the training input, UCS activates the training phase to produce the solution. For the test input, it activates only the test phase in order to evaluate the system classification accuracy as the UCS performance. In addition, we evaluate the population size (i.e., the number of rules in $[P]$) to evaluate the generalization capacity of UCS. The UCS performance and the population size are reported as an average of 30 trials. We test UCS and UCS with our rule-repair algorithm (denoted by "Ours").

We use the following UCS parameter settings with respect to [9, 12]; $\beta = 0.2$, $\delta = 0.1$, $\nu = 10$, $\theta_{GA} = 25$, $\chi = 0.8$, $\theta_{\mathrm{del}} = 20$, $\theta_{\mathrm{sub}} = 20$, $acc_0 = 0.99$, $P_{\#} = 0.8$, $\mu = 0.04, \tau = 0.4$, $s_0 = 1.0$, and $m_0 = 0.1$. The subsumptions are turned on. For $\{6, 11, 20, 37\}$-RMUXs and $\{11, 15\}$-RMOPs, we set N and the maximum iterations to $\{800, 5000, 30000, 30000, 10000, 100000\}$ and $\{50000, 200000, 1000000, 1500000, 200000, 500000\}$, respectively. For our repair algorithm, we set $D = 0.01$.

4.3   Result

Figures 2 and 3 summarize the performances and the population size. As shown in Fig. 2, our rule-repair algorithm successfully boosts the UCS performance on all the problems employed in this paper. Specifically, our rule-repair algorithm improves the performance at early iterations, where we can expect that many inaccurate rules exist in $[!C]$. For instance, it reaches almost the optimal performance after 100,000 iterations on 11-RMUX, but UCS requires 200,000 iterations to reach it. The performances on RMOPs do not reach the optimal performance due to the complexity of the overlapping problem, which is a similar tendency to existing works [25, 29].

A possible drawback of our rule-repair algorithm is to increase the population size since it enhances a bias to produce specific rules with less rule-generality. As shown in Fig. 3, this insight can be observed in early generations. However, our rule-repair algorithm successfully prevents the increase of the population size over iterations. Besides, it produces a more compact population than that of UCS on 6, 11-RMUXs. Note that the population size reaches the maximum population size N if the cover-delete cycle occurs. This tendency can be observed for both UCSs on RMOPs, and so our rule-repair algorithm itself does not provoke the cover-delete cycle. We suspect that N should be further increased to cover various niches defined in RMOPSs [29]. Thus, we can empirically confirm that our rule-repair algorithm successfully prevents the cover-delete cycle.

Finally, we further give empirical insights to confirm the efficiency of our rule-repair algorithm. Figure 4 shows the summation of numerosity of rules in $[!C]$ over iterations (i.e., $\sum_{cl \in [!C]} cl.num$), on 11-RMUX and 11-RMOP. Note that the population involves many inaccurate rules if $\sum_{cl \in [!C]} cl.num$ is a large value since inaccurate rules tend to be over-generalized; and so those rules frequently match inputs, resulting in the increase of the $\sum_{cl \in [!C]} cl.num$. From the figure, it is obvious that our rule-repair algorithm contributes to decreasing the summation of numerosity of $[!C]$. This means that our rule-repair algorithm successfully reduces the inaccurate rules by improving their classification accuracy.

In summary, our rule-repair algorithm successfully repairs the inaccurate rules, and thus it boosts the UCS performance while preventing the increase of the population size as well as the cover-delete cycle.

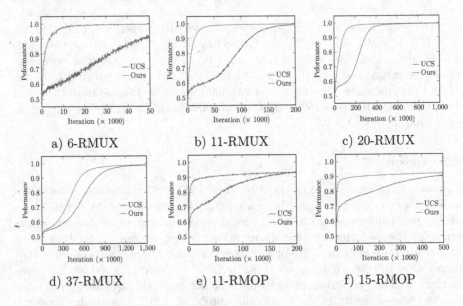

Fig. 2. The performances of UCS and the proposal.

5   Analysis

This section presents analytical insights into our minimum rule-repair strategy.

5.1   Analysis of the Number of Sub-conditions to Be Repaired

One of our strategy to minimize the reduction of the rule-generality is to repair either l_i or u_i for *one* sub-condition c_i. In this subsection, we validate the impact of this strategy. In detail, we here extend our algorithm to repair Ψ sub-conditions at the same time, where Ψ is the number of sub-conditions to be repaired; and a default setting is $\Psi = 1$. More specifically, we insert a "for" loop (i.e., for $i = 1$ to Ψ) between line 3 and 4 in Algorithm 1; and each dimension index k is randomly sampled with no duplicates.

We here compare the performances of UCS with the rule-repair algorithm for $\Psi = \{1, 3, 5\}$. Note that the rule-generality of repaired rules should be rapidly decreasing with the increase of Ψ. Figures 5 and 6 show the performances and the population size on 11, 20, 37-RMUXs with the same experimental settings (see Subsect. 4.2), respectively. As shown in those figures, the performance gradually degrades with the increase of Ψ. This tendency is clearly highlighted with the increase of the problem dimensions $d = \{11, 20, 37\}$. Besides, the population size

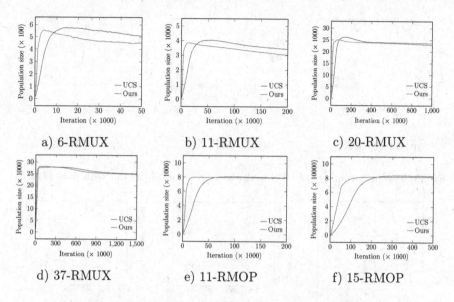

a) 6-RMUX   b) 11-RMUX   c) 20-RMUX

d) 37-RMUX   e) 11-RMOP   f) 15-RMOP

Fig. 3. The population sizes of UCS and the proposal.

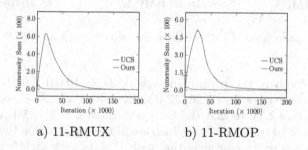

a) 11-RMUX   b) 11-RMOP

Fig. 4. The summation of numerosity of rules in $[!C]$.

with $\Psi = 1$ decreases slightly faster than $\Psi = \{3, 5\}$. This indicates that our minimum rule-repair strategy (i.e., $\Psi = 1$) successfully discovers the optimum rules faster than the other settings in this paper. Accordingly, those experimental results empirically confirm our hypothetical insight; the rule-condition should be repaired with as minimum reductions of the rule generality as possible.

5.2 Analysis of the Minimum Distance D

Next, we analyze the impact of the hyperparameter D, which controls the minimum distance between a specific input value x_k and $\hat{l}_k$ (or $\hat{u}_k$). As noted in Sect. 3, the reduction of the rule-generality can be also minimized in terms of D, e.g., $D = 10^{-10}$. Thus, according to our hypothesis (i.e., to minimize the reduction of the rule-generality), D may be an important parameter dependent on the performance.

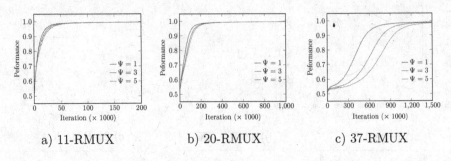

a) 11-RMUX   b) 20-RMUX   c) 37-RMUX

Fig. 5. The performances of UCS with the rule-repair algorithm with $\Psi = \{1, 3, 5\}$

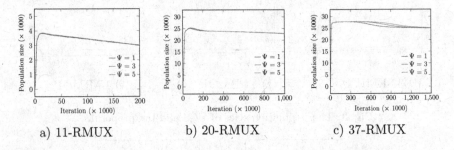

a) 11-RMUX   b) 20-RMUX   c) 37-RMUX

Fig. 6. The population of UCS with the rule-repair algorithm with $\Psi = \{1, 3, 5\}$

We here compare the performances with $D = \{0.001, 0.01, 0.1\}$. Figure 7 shows the performances on 11, 20, 37-RMUXs with the same experimental settings as in Sect. 4.2. Note that the UCS performance of 37-RMUX is reported as an average of 10 trials due to the increase of the computation time. As shown in this figure, our rule-repair algorithm with $D = 0.1$ very slightly improves the performance; however, compared to the impact of Ψ, no significant impact of D is observed even with the increase of the problem dimensions. This is because that D does not significantly change a matching probability of the rule-condition to an input. For instance, suppose 11-MUX and a rule-condition C with $\{l_i = 0.1, u_i = 0.9\}, \forall i = \{1, 2, \cdots, 11\}$, its matching probability can be $0.0859 (= (0.9 - 0.1)^{11})$ for randomly-sampled inputs under a uniform distribution. Then, consider that we repair l_1 with the first element of input $x_i = 0.4$, the sub-condition c_1 can be rewritten as $\{l_1 = 0.4 + D, u_1 = 0.9\} (i = 1)$. Then, the matching probability of C can be $\{0.0536, 0.0526, 0.0430\}$ for $D = \{0.001, 0.01, 0.1\}$, respectively. Consequently, we suppose that D cannot be an important parameter dependent on the performance unless it is set to an enough small value.

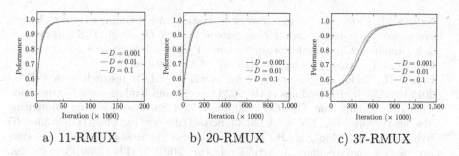

a) 11-RMUX   b) 20-RMUX   c) 37-RMUX

Fig. 7. The performances of UCS with the rule-repair algorithm with $D = \{0.001, 0.01, 0.1\}$

6   Conclusion

This paper proposed Minimum Rule-repair Algorithm (MRA) for UCS with real-valued inputs on classification problems. Our concept is to repair inaccurate rules with a possible minimum reduction of the rule-generality in order to avoid the problematic cover-delete cycle. Accordingly, we identified the following two principles to achieve this purpose; 1) to repair rule-condition to eliminate one incorrect input from a matching subspace represented by its rule-condition, and 2) to repair either a lower value or an upper value for one dimension x_i. Experimental results confirmed the adequacy of those principles. Consequently, UCS with MRA successfully boosts the performance while preventing the increase of the population size as well as the cover-delete cycle.

In future work, we apply our algorithm to real-world classification tasks, where some difficulties, e.g., missing attribute and class imbalance, can be observed. We further analyze the impact of the hyperparameter D (the minimum distance) on non-uniformed distributions of inputs, aiming to reveal its optimal setting up guide.

References

1. Holland, J.H.: Escaping brittleness: the possibilities of general purpose learning algorithms applied to parallel rule-based systems. Mach. Learn. Artif. Intell. Approach **2**, 593–623 (1986)
2. Kovacs, T.: Evolving optimal populations with XCS classifier systems. Technical report CSR-96-17 and CSRP-96-17, School of Computer Science, University of Birmingham, Birmingham, U.K. (1996)
3. Butz, M.V., Goldberg, D.E., Tharakunnel, K.: Analysis and improvement of fitness exploitation in XCS: bounding models, tournament selection, and bilateral accuracy. Evol. Comput. **11**(3), 239–277 (2003)
4. Borna, K., Hoseini, S., Aghaei, M.A.M.: Customer satisfaction prediction with Michigan-style learning classifier system. SN Appl. Sci. **1**(11), 1450 (2019)

5. Nakata, M., Chiba, K.: Design strategy generation for a sounding hybrid rocket via evolutionary rule-based data mining system. In: Leu, G., Singh, H.K., Elsayed, S. (eds.) Intelligent and Evolutionary Systems. PALO, vol. 8, pp. 305–318. Springer, Cham (2017). https://doi.org/10.1007/978-3-319-49049-6_22

6. Pätzel, D., Stein, A., Nakata, M.: An overview of LCS research from IWLCS 2019 to 2020. In: Proceedings of the 2020 Genetic and Evolutionary Computation Conference Companion, GECCO 2020, pp. 1782–1788. Association for Computing Machinery, New York, NY, USA (2020). https://doi.org/10.1145/3377929.3398105

7. Urbanowicz, R.J., Moore, J.H.: Learning classifier systems: a complete introduction, review, and roadmap. J. Artif. Evol. App. **2009**, 1–1125 (2009). https://doi.org/10.1155/2009/736398

8. Wilson, S.W.: Classifier fitness based on accuracy. Evol. Comput. **3**(2), 149–175 (1995)

9. Bernadó-Mansilla, E., Garrell-Guiu, J.M.: Accuracy-based learning classifier systems: models, analysis and applications to classification tasks. Evol. Comput. **11**(3), 209–238 (2003)

10. Roozegar, M., Mahjoob, M., Esfandyari, M., Panahi, M.S.: XCS-based reinforcement learning algorithm for motion planning of a spherical mobile robot. Appl. Intell. **45**(3), 736–746 (2016)

11. Ebadi, T., Zhang, M., Browne, W.: XCS-based versus UCS-based feature pattern classification system. In: Proceedings of the 14th Annual Conference on Genetic and Evolutionary Computation, pp. 839–846 (2012)

12. Nakata, M., Browne, W.N.: Learning optimality theory for accuracy based learning classifier systems. IEEE Trans. Evol. Comput. **25**(1), 61–74 (2020)

13. Matsumoto, K., Takano, R., Tatsumi, T., Sato, H., Kovacs, T., Takadama, K.: XCSR based on compressed input by deep neural network for high dimensional data. In: Proceedings of the Genetic and Evolutionary Computation Conference Companion, pp. 1418–1425 (2018)

14. Tadokoro, M., Hasegawa, S., Tatsumi, T., Sato, H., Takadama, K.: Knowledge extraction from XCSR based on dimensionality reduction and deep generative models. In: 2019 IEEE Congress on Evolutionary Computation (CEC), pp. 1883–1890. IEEE (2019)

15. Debie, E., Shafi, K.: Implications of the curse of dimensionality for supervised learning classifier systems: theoretical and empirical analyses. Pattern Anal. Appl. **22**(2), 519–536 (2019)

16. Debie, E., Shafi, K., Lokan, C., Merrick, K.: Reduct based ensemble of learning classifier system for real-valued classification problems. In: 2013 IEEE Symposium on Computational Intelligence and Ensemble Learning (CIEL), pp. 66–73. IEEE (2013)

17. Urbanowicz, R.J., Moore, J.H.: ExSTraCS 20: description and evaluation of a scalable learning classifier system. Evol. Intell. **8**(2), 89–116 (2015)

18. Aenugu, S., Spector, L.: Lexicase selection in learning classifier systems. In: Proceedings of the Genetic and Evolutionary Computation Conference, pp. 356–364 (2019)

19. Moschoyiannis, S., Shcherbinin, V.: Fine tuning run parameter values in rule-based machine learning. In: RuleML+ RR (Supplement) (2019)

20. Abedini, M., Kirley, M.: Guided rule discovery in XCS for high-dimensional classification problems. In: Wang, D., Reynolds, M. (eds.) AI 2011. LNCS (LNAI), vol. 7106, pp. 1–10. Springer, Heidelberg (2011). https://doi.org/10.1007/978-3-642-25832-9_1

21. Abedini, M., Kirley, M.: An enhanced XCS rule discovery module using feature ranking. Int. J. Mach. Learn. Cybern. **4**(3), 173–187 (2013)
22. Butz, M.V., Kovacs, T., Lanzi, P.L., Wilson, S.W.: Toward a theory of generalization and learning in XCS. IEEE Trans. Evol. Comput. **8**(1), 28–46 (2004)
23. Butz, M.V., Goldberg, D.E., Lanzi, P.L.: Computational complexity of the XCS classifier system. In: Bull, L., Kovacs, T. (eds.) Foundations of Learning Classifier Systems. Studies in Fuzziness and Soft Computing, vol. 183, pp. 91–125. Springer, Heidelberg (2005). https://doi.org/10.1007/11319122_5
24. Lanzi, P.L., et al.: A study of the generalization capabilities of XCS. In: ICGA, pp. 418–425. Citeseer (1997)
25. Iqbal, M., Browne, W.N., Zhang, M.: Reusing building blocks of extracted knowledge to solve complex, large-scale boolean problems. IEEE Trans. Evol. Comput. **18**(4), 465–480 (2013)
26. Tadokoro, M., Hasegawa, S., Tatsumi, T., Sato, H., Takadama, K.: Local covering: adaptive rule generation method using existing rules for XCS. In: 2020 IEEE Congress on Evolutionary Computation (CEC), pp. 1–8. IEEE (2020)
27. Wilson, S.W.: Mining oblique data with XCS. In: Luca Lanzi, P., Stolzmann, W., Wilson, S.W. (eds.) IWLCS 2000. LNCS (LNAI), vol. 1996, pp. 158–174. Springer, Heidelberg (2001). https://doi.org/10.1007/3-540-44640-0_11
28. Wilson, S.W.: Get real! XCS with continuous-valued inputs. In: Lanzi, P.L., Stolzmann, W., Wilson, S.W. (eds.) IWLCS 1999. LNCS (LNAI), vol. 1813, pp. 209–219. Springer, Heidelberg (2000). https://doi.org/10.1007/3-540-45027-0_11
29. Nakata, M., Browne, W., Hamagami, T., Takadama, K.: Theoretical XCS parameter settings of learning accurate classifiers. In: Proceedings of the Genetic and Evolutionary Computation Conference. GECCO 2017, pp. 473–480. Association for Computing Machinery, New York, NY, USA (2017). https://doi.org/10.1145/3071178.3071200

Applications

A Biased Random-Key Genetic Algorithm for the 2-Dimensional Guillotine Cutting Stock Problem with Stack Constraints

Marcos V. A. Guimarães[1]([✉]), Eduardo T. Bogue[2], Iago A. Carvalho[1], Armando H. Pereira[1], Thiago F. Noronha[1], and Sebastián Urrutia[3]

[1] Department of Computer Science, Universidade Federal de Minas Gerais, Belo Horizonte, MG, Brazil
mvaguimaraes@gmail.com, tfn@dcc.ufmg.br
[2] Universidade Federal de Mato Grosso do Sul, Campo Grande, Brazil
[3] Molde University College, Molde, Norway

Abstract. This paper tackles the 2-Dimensional Guillotine Cutting Stock Problem with Stack Constraints. The problem asks for the cutting of a set of items with the minimum amount of raw material. The cutting patterns are subject to a number of constraints, including a new realistic constraint, regarding item precedence, which has just been introduced in the literature. In this case, the items are organized in stacks, where each stack represents a customer request and defines the order in which the items must be cut. That is, if item i precedes item j within a stack, then i must be cut before j. However, there is no precedence constraint between items in different stacks. This constraint comes from applications where items must be stacked and shipped in the exact order that they will be used by the customer, thus avoiding the risk of damaging fragile items (as is the case in the glass industry) or the cost of moving heavy items (as is the case in the steel industry). We propose two heuristics, one Evolutionary Algorithm (EA) adapted from a similar problem in the literature, and a novel Biased Random Key Genetic Algorithm (BRKGA). Computational results show that BRKGA outperforms the evolutionary algorithm from the literature.

Keywords: Cutting stock · Dynamic programming · Evolutionary algorithm · Guillotine cut · Stack constraints

1 Introduction

The problem of cutting a large plate of raw material into a specified set of smaller objects is a common industrial challenge in glass [1], paper [2], wood [3], and steel [4] industries, among others. This problem is referred as the 2-dimensional cutting stock problem (2DCSP for short) [5]. It aims at cutting all the smaller objects with the minimum amount of raw material. In this paper, we refer to

© Springer Nature Switzerland AG 2022
B. Dorronsoro et al. (Eds.): META 2021, CCIS 1541, pp. 155–169, 2022.
https://doi.org/10.1007/978-3-030-94216-8_12

the large plate of raw material simply as *plate* and to the smaller objects as *items*. Besides, as in [6–11], we assume that they are two-dimensional and have a rectangular shape.

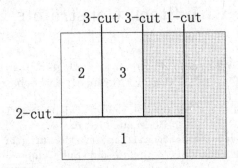

Fig. 1. An example of a cutting pattern.

We focus on the variant of 2DCSP called the 2-dimensional 3-staged cutting stock problem subject to guillotine constraints (2DCSP-3S for short) [6,7,9,10]. In this variant, only *guillotine cuts* are allowed, *i.e.*, cuts that go from one side to the opposite side of the plate and split it into two rectangular pieces. These cuts are divided into *stages*, where each cutting stage consists of a sequence of parallel guillotine cuts. At each stage, the cuts are orthogonal to those of the previous stage, since each piece of plate is rotated by 90° before the next cutting stage begins. We refer to a k-cut as a guillotine cut performed in the k-th stage. Besides, we assume that the plate is oriented such that its width is larger than its height, and that the odd staged cuts are vertically oriented, while the even staged cuts are horizontally oriented. An example of a 3-stage cutting pattern used to separate three items from a plate is shown in Fig. 1, where the items are numbered from 1 to 3 and the unused pieces of plate are shadowed.

This cutting pattern can be interpreted as a tree, where the root node (at level 0) corresponds to the whole plate, and each node in the k-th level of the tree corresponds to a piece of plate obtained from a k-cut to the piece of plate of its parent node. Therefore, the leaves of this tree corresponds to either items or unused pieces of plate. It is assumed that cuts are performed using a depth first approach in this tree to avoid changing the piece of plate that is in the guillotine. An example of the tree representing the cutting pattern of Fig. 1 is given in Fig. 2. First, a vertical 1-cut is applied to the root node to detach an unused piece of plate. Next, a horizontal 2-cut is performed to extract the item 1. Then, two successive vertical 3-cuts are executed to obtain items 2 and 3, as well as another unused piece of plate.

In the case of 2DCSP-3S, the maximum number of stages is limited to three. However, a single additional 4-cut is allowed if and only if it is used to separate a single item from an unused piece of plate [6,12–14]. This is known in the literature as *trimming*. Figure 3a shows an example of trimming, where the item

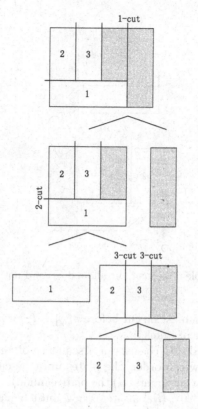

Fig. 2. Representation of the cutting pattern of Fig. 1 as a tree.

3 is separated from an unused piece of plate by a single 4-cut, while Fig. 3b gives an example of an invalid 4-cut used to separate item 3 from item 4.

In this paper, we deal with a variant of 2DCSP-3S that has additional precedence constraints that was recently introduced in [15]. In this case, the items are organized in stacks, where each stack represents a customer request and defines the order in which the items must be cut. That is, if item i precedes item j within a stack, then i must be cut before j. However, there is no precedence constraint between items in different stacks. This constraint comes from applications where items must be stacked and shipped in the exact order that they will be used by the customer, thus avoiding the risk of damaging fragile items (as is the case in the glass industry) or the cost of moving heavy items (as is the case in the steel industry). We refer to this variant of 2DCSP-3S as the 2-dimensional Guillotine Cutting Stock Problem with Stack Constraints (2DCSP-SC). This problem is formally defined below.

Let W and H be the width and the height of the plates, respectively, and I be the set of items to be cut, where each item $i \in I$ has height h_i and width w_i. Besides, let S be the set of stacks that represent customer orders, whereas $b = (\pi_1^s, \pi_2^s, \ldots, \pi_{n_s}^s)$ describes the order in which the items from stack $s \in S$ must

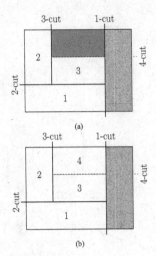

Fig. 3. Example of a *4-cut* cut allowed (a) and not allowed (b).

be cut, such that $\pi_k^s \in I$ must be cut before $\pi_{k+1}^s \in I$, for all $k = \{1,\ldots,n_s-1\}$, where n_s is the number of items in s.

A solution to 2DCSP-SC consists of a sequence of cutting patterns P that describes how, and in which order, the plates must be cut. This solution must satisfy all the following constraints: (i) the plate cannot be rotated; (ii) the items can only be rotated by $90°$; (iii) all items in I must be cut exactly once; (iv) if $i \in I$ precedes $j \in I$ in a stack, then i must be cut before j; (v) only guillotine cuts are allowed; and (vi) the number of cutting stages is at most three along with the additional 4-cut, as previously described.

The cost $f(P)$ of a solution P to 2DCSP-SC is the amount of raw material used to cut all items. As in [6,16–18], the unused pieces of plate that result from the cutting patterns P are divided into two types. The so called *leftover* is the material to the right of the last 1-cut applied to the last plate. It is assumed that this piece of plate can be reused, and it is not considered in $f(P)$. All the other unused pieces of plate are considered *waste*, as it is assumed that they cannot be reused. For example, in Fig. 3a, the unused piece of plate colored in gray is considered waste, while that filled with dots is considered leftover. The objective function $f(P)$ is defined as

$$f(P) = H \cdot W \cdot (|P| - 1) + H \cdot r(P),$$

where $|P|$ denotes the number of plates used, $H \cdot W \cdot (|P| - 1)$ is the total area of the first $|P| - 1$ plates. Furthermore, $r(P)$ gives the position of the last *1-cut* on the last cutting pattern of P. Thus, $H \cdot r(P)$ represents the used area of the last plate.

Let Δ be the set of feasible solutions for 2DCSP-SC. This problem consists of finding a solution $P^* = \mathrm{argmin}_{P \in \Delta} f(P)$, *i.e.*, the cutting patterns that use the least amount of raw material to cut all items in I. When there is only one item per stack, this problem reduces to 2DCSP-3S. Since 2DCSP-3S is NP-Hard [19], 2DCSP-SC is also NP-Hard.

As there is no known technique to design a polynomial-time exact algorithm for NP-Hard problems, this paper focus on heuristic algorithms. However, as far as we can tell, the new precedence constraints introduced in [15] preclude the use of most algorithms in the literature related to 2DCSP-3S [13,16,17,20], except those in [21], because they were not designed to consider item precedence. Therefore, this paper adapts an Evolutionary Algorithm (EA) described in [21], and also proposes a Biased Random-key Genetic Algorithm (BRKGA) to address the 2DCSP-SC. Computational experiments show that BRKGA outperforms the EA of the literature.

The remainder of this paper is organized as follows. First, related work are discussed in Sect. 2. Next, a constructive algorithm for 2DCSP-SC, which is used as a decoder for the other heuristics, is proposed in Sect. 3. Then, the adaptation to 2DCSP-SC of the Evolutionary Algorithm of [21] is detailed in Sect. 4, and the proposed Biased Random-key Genetic Algorithm is described in Sect. 5. Finally, computational experiments are reported in Sect. 6 and concluding remarks are drawn in the last section.

2   Related Work

A Sequential Heuristic Procedure (SHP) was proposed in [22]. The first stage of this heuristic selects the height of the cut, the second stage the length of the cut, and the third stage the number of times the generated cut pattern will be used. The authors concluded that the performance of the proposed heuristic is better than heuristics that use fixed measures to define the sizes of the cuts. A variant of 2DCSP-3S in which the plates may contain defects and vary in size was addressed in [23]. The proposed heuristic first sorts the larger sides of the plates in ascending order, and the items are sorted in decreasing order following the same criteria. The algorithm tries to position the widest items on the smallest plates and after all the items are positioned, it checks if any item was placed in any defective area; if so, that item is removed, and the possibility of being added to any of the plates already used is verified. Computational experiments showed that these heuristics obtained better results than those presented in [24].

A heuristic procedure based on Variable Neighborhood Search (VNS) was proposed in [25]. To build an initial solution, three heuristics based on the first-fit approach of [26] were used (3-staged First Fit Decreasing Height with rotations, matching step and Fill Strip), so that the best solution provided by them is selected. Computational experiments concluded that the heuristic proposed in this work provided better solutions than the VNS approach present in [26].

A heuristic that combines a recursive approach and a Beam Search algorithm was proposed in [17]. Unlike branch and bound algorithms, in Beam Search, only elite nodes with high potential are investigated [27]. In this approach, the recursion is used to generate segments of strips, and a Beam Search heuristic is used to obtain the 3-staged cutting patterns considering usable leftover. Computational experiments showed that the heuristic proposed in this work obtained better solutions than those of [6].

A Finite First Fit Heuristic (FFF), an Evolutionary Algorithm (EA), and two strategies based on branch-and-bound have been proposed to solve 2DCSP-3S in [21]. Computational experiments showed that EA obtained better results than the other heuristics. As far as we know, FFF and EA are the best heuristics in the literature that can be adapted to handle the precedence constraint of 2DCSP-SC. Therefore, they are adapted to 2DCSP-SC in Sects. 3 and 4, respectively.

3   Finite First Fit (FFF) Heuristic

In this section, we extend the FFF heuristic of [21] to 2DCSP-SC. This heuristic assumes that all the items are oriented in such a way that their width is greater than or equal to their height, and it does not perform any further rotations. It starts with an empty solution, and, at each iteration, it inserts an item in the solution using the first fit approach described in Algorithm 1.

FFF inserts the items in the order they appear in a permutation Π of the items in I. This permutation is such that if an item i precedes an item j in any stack, i precedes j in Π. This property is necessary to guarantee that there is always a place to insert an item without violating the precedence constraints. That is the case because when an item is inserted in the solution, all preceding items have already been inserted. In the worst case scenario, the next item can be inserted in a new empty plate. The sorting algorithm used to generate permutations that satisfy this property is explained in the next section.

Instead of describing Algorithm 1 using the usual recursive tree representation of a solution, we adopt a novel representation based on what we called a *k-box* representation. A 0-box by definition is a plate. Therefore, it has always height H and width W. Such a box is divided into a sequence of 1-boxes. Therefore, a 1-box has always height H, but can have variable width. It is assumed that the 1-boxes are placed contiguously from the left to the right of their respective 0-box. Therefore, all the area of the 0-box not covered by a 1-box is unused area (waste or leftover). Analogously, a 1-box is divided into a sequence of 2-boxes. Therefore, a 2-box has always the same width of its respective 1-box, but can have variable height. It is assumed that the 2-boxes are placed contiguously from the bottom to the top of their respective 1-box. Therefore, all the area of the respective 1-box not covered by a 2-box is waste. Similarly, a 2-box is divided into a sequence of 3-boxes, each one associated to exactly one item. Therefore, a 3-box has always the same height of its respective 2-box, but its width is exactly the same as that of its corresponding item. It is worth pointing out that, from the width of the 3-box, one can infer if the corresponding item was rotated or not. Besides, it is implicit that if the height of a 3-box is larger than that of the corresponding item, a trimming 4-cut occurs. It is also assumed that the 3-boxes are placed contiguously from the left to the right of their respective 2-boxes. Therefore, all the area of the respective 2-cut not covered by a 3-box is waste. It can be observed that there is a direct correspondence between a

k-box and a level k node in the cutting pattern tree. The advantage of the k-box representation is that one does not need to account to the exact position of the nested k-cuts, but only to the size of the corresponding k-boxes, which can be inferred from the width and height of the items in each box.

The pseudo-code of FFF is described in Algorithm 1. This heuristic receives an instance $\mathcal{I}$ of 2DCSP-SC and a permutation Π of the items in I, and returns a solution corresponding to a sequence of cutting patterns P that describes how to cut all the items in I without breaking the precedence constraints. In line 1, an empty partial solution P is initialized. At each iteration of the loop of lines 2 to 22, the next item i, according to the permutation Π, is inserted in the solution. Next, at each iteration of the loop of lines 3 to 20, FFF scans every 0-box b^0 in the order they appear in P. Then, at each iteration of the loop of lines 4 to 15, FFF evaluates every 1-box b^1 in the order they appear in b^0. Following, at each iteration of the loop of lines 5 to 10, FFF inspects every 2-box b^2 in the order they appear in b^1. If the item i fits in b^2 (see line 6), a new 3-box with i is initialized and is appended to b^2 in line 7, and the heuristic continues to the next item in Π in line 8. It is worth noting that only the 2-boxes whose items are cut after those that precede i are considered. Otherwise, i could be cut before an item that precedes it. On the other hand, if i does not fit in b^2 but it fits as

---

Algorithm 1. *Finite First-Fit* (FFF)

Input: $\mathcal{I}$ and Π
Output: P

1: $P \leftarrow [\,]$
2: **for each** i in Π **do**
3:      **for each** b^0 in P **do**
4:          **for each** b^1 in b^0 **do**
5:              **for each** b^2 in b^1 **do**
6:                  **if** i fits in b^2 respecting the precedence constraint **then**
7:                      $b^3 \leftarrow i$, and $b^2 \leftarrow b^2 : b^3$
8:                      **continue** to the next item in Π
9:                  **end if**
10:             **end for**
11:             **if** $[i]$ fits in b^1 respecting the precedence constraint **then**
12:                 $b^2 \leftarrow [i]$, and $b^1 \leftarrow b^1 : b^2$
13:                 **continue** to the next item in Π
14:             **end if**
15:         **end for**
16:         **if** $[[i]]$ fits in b^0 respecting the precedence constraint **then**
17:             $b^1 \leftarrow [[i]]$, and $b^0 \leftarrow b^0 : b^1$
18:             **continue** to the next item in Π
19:         **end if**
20:     **end for**
21:     $b^0 \leftarrow [[[i]]]$, and $P \leftarrow P : b^0$
22: **end for**
23: **return** P

---

a new 2-box on top of b^1 without breaking the precedence constraint (see line 11), a new 2-box (containing a single 3-box with i) is appended to b^1 in line 12, and FFF continues to the next item in Π in line 13. Moreover, if the latter is not possible, but i fits as a new 1-box to the right of b^0 without breaking the precedence constraint (see line 16), a new 1-box (containing a single 2-box with i) is appended to b^0 in line 17, and FFF continues to the next item in Π in line 18. Finally, if i does not fit any box of the current solution, a new 0-box (plate) is appended to P in line 21 with a single 1-box containing i. When all the items are inserted in P the solution is returned in line 23. This heuristic is used as a decoder in both the Evolutionary Algorithm of [21] and the Biased Random-key Genetic Algorithm introduced in Sects. 4 and 5, respectively.

4   An Evolutionary Algorithm for 2DCSP-SC

In this section, the Evolutionary Algorithm (EA) of [21] is adapted to address 2DCSP-SC. This algorithm represents a solution as an $|I|$-vector, in which each component is a real number (referred to as *key*) in the range $[0, 1]$ associated with an item in I. Each solution is decoded by a decoding heuristic that receives the vector of keys and builds a feasible solution for 2DCSP-SC. Let k_i be the key associated with the item $i \in I$ and w_i be the width of i, the decoding of EA consists of two steps. First, a permutation Π is generated accordingly to the following selection sort algorithm. Let $\rho_i = k_i \cdot w_i$, at each iteration of the sorting algorithm, the item with the largest value of ρ_i, that is on top of a stack in S, is popped from its stack and added to the end of Π. Then, the FFF heuristic is run with the resulting permutation of items. The cost of the solution returned by FFF is used as the fitness of the chromosome.

EA is a steady-state evolutionary algorithm, where each offspring replaces the worst solution in the population. Initial solutions are created at random, and at each iteration two parent solutions are selected randomly. Then, the Order 3 Crossover (OX3) of [28] is applied to these solutions to generate a new offspring. Two mutation operators are used: (i) the Reciprocal Exchange (RX), which chooses two items at random and swaps their keys; and (ii) the Block Exchange (BX), which swaps the keys of two non-overlapping blocks of consecutive items. The size of these blocks is set to $\lceil 2^R \rceil$, as suggested by [21], where R is a random value in the interval $(0, \lfloor ld\frac{n}{2} \rfloor]$, in order to allow shorter blocks to be chosen more likely. The number of mutations applied to each new offspring solution is chosen as a Poisson-distributed random variable with expected value 2. Every time a mutation is applied to the offspring, either RX or BX is randomly chosen with equal probability.

5   Biased Random-Key Genetic Algorithm

Random-key Genetic Algorithms (RKGA) were first introduced by Bean [29] for combinatorial optimization problems for which solutions can be represented as a permutation vector. In this approach, two parents are selected at random from the entire population to implement the crossover operation in the implementation of a RKGA. Parents are allowed to be selected for mating more than once in a given generation.

A Biased Random-key Genetic Algorithm (BRKGA) differs from a RKGA in the way parents are selected for crossover, see Gonçalves and Resende [30] for a review. In a BRKGA, each element is generated combining one element selected at random from the elite solutions in the current population, while the other is a non-elite solution. We say the selection is biased since one parent is always an elite individual and because this elite solution has a higher probability of passing its genes to the offsprings, *i.e.*, to the new generation. A BRKGA provides a better implementation of the essence of Darwin's principle of "survival of the fittest" than the RKGA, since an elite solution has a higher probability of being selected for mating and the offsprings have a higher probability of inheriting the genes of the elite parent.

The BRKGA for 2DCSP-SC evolves a population of chromosomes that consists of $|I|$-vectors of keys, which are decoded exactly as in the Evolutionary Algorithm described in Sect. 4. We use the parameterized uniform crossover scheme proposed in [31] to combine two parent solutions and produce an offspring. In this scheme, the offspring inherits each of its keys from the best fit of the two parents with probability $\rho > 0.5$ and from the least fit parent with probability $1 - \rho$. BRKGA do not make use of the standard mutation operator, where parts of the chromosomes are changed with a small probability. Instead, the following concept of mutants is used: a fixed number of mutant solutions are introduced in the population in each generation, randomly generated in the same way as in the initial population. Mutants play the same role of the mutation operator in traditional genetic algorithms, diversifying the search and helping the procedure to escape from locally optimal solutions.

The keys associated to each item are randomly generated in the initial population. At each generation, the population is partitioned into two sets: *TOP* and *REST*. Consequently, the size of the population is $|TOP| + |REST|$. Subset *TOP* contains the best solutions in the population. Subset *REST* is formed by two disjoint subsets: *MID* and *BOT*, with subset *BOT* being formed by the worst elements on the current population. As illustrated in Fig. 4, the chromosomes in *TOP* are simply copied to the population of the next generation. The elements in *BOT* are replaced by newly created mutants that are placed in the new set *BOT*. The remaining elements of the new population are obtained by crossover with one parent randomly chosen from *TOP* and the other from *REST*. This distinguishes a biased random-key GA from the random-key genetic algorithm of Bean [29] (where both parents are selected at random from the entire population). Since a parent solution can be chosen for crossover more than once in a given generation, elite solutions have a higher probability of passing their random keys to the next generation. In this way, $|MID| = |REST| - |BOT|$ offspring solutions are created. The algorithm stops when a maximum elapsed time is reached.

6   Computational Experiments

The computational experiments reported in this section evaluates the performance of the BRKGA and the EA heuristics. These algorithms were implemented in C++ and compiled with GNU *gcc* version 6.3. The population size of both heuristics was set to 1000 solutions, and the stopping criteria was set to 10 min of running time. All experiments were performed in a single core of an Intel Xeon machine with 2.00 GHz of clock speed and 16 GB of RAM. As both BRKGA and EA relies on stochastic operators, we ran these heuristics 20 times for each instance using different seeds for the Mersenne Twister pseudo-random number generator [32].

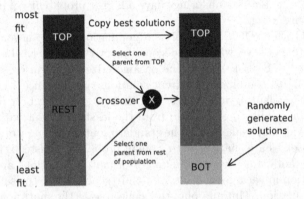

Fig. 4. Population evolution between consecutive generations of a BRKGA.

Three sets of instances were used in the experiments, namely *Set A*, *Set B*, and *Set X*. These instances were adapted from realistic ones employed in the ROADEF Challenge 2018, which tackled a similar problem commissioned by Saint-Gobain Glass France, which is one of the world's leading flat glass manufacturers. The original instances resemble scenarios found in Saint-Gobain factories and can be retrieved from the website http://www.roadef.org/challenge/2018/en/instances.php. We adapted these instances by ignoring the data regarding specific constraints of Saint-Gobain's guillotines, and using only the data necessary for 2DCSP-SC. In every instance, the plates have $W = 6000$ and $H = 3210$. Table 1 presents the characteristics of each instance set. One can see that the instances greatly vary. The number of stacks range from 1 to 247, while the number of items stretch from 5 to 656. Furthermore, the smallest width of an item is only 345, while the maximum width of an item is equal to 3495. Additionally, the height of the items also vary between 123 and 2010.

The results for this experiment are displayed in Tables 2, 3 and 4, whereas each table gives the result for a different instance set. The first column of each table displays the instance name, while the second column presents a lower bound (LB) computed as $\sum_{i \in I} w_i h_i$, *i.e.*, the sum of the area of all items in I. The

Table 1. Characteristics of each instance set

	Set A	Set B	Set X
Number of instances	20	15	15
Min. number of items	5	68	124
Ave. number of items	101.95	303.87	284.33
Max. number of items	392	656	412
Min. number of stacks	1	2	2
Ave. number of stacks	11.20	32.00	28.53
Max. number of stacks	72	241	247
Min. item width	345	351	353
Ave. item width	1317.35	1410.69	1317.40
Max. item width	3495	2952	2813
Min. item height	137	123	193
Ave. item height	594.82	668.63	600.03
Max. item height	2010	1759	1828

Table 2. Results for the Set A of instances

Instance	LB	BRKGA		EA	
		gap (%)	cv (%)	gap (%)	cv (%)
A1	4514704	10.87	0.00	10.87	0.00
A2	77201851	13.32	1.74	13.64	0.92
A3	41796990	19.79	1.10	20.80	1.19
A4	41796990	19.90	0.92	21.06	1.17
A5	56570007	15.55	1.25	17.77	1.16
A6	43254870	14.40	0.82	15.81	0.11
A7	70195170	20.02	1.16	22.25	1.17
A8	138045196	20.53	1.92	22.45	0.80
A9	44879034	21.47	4.63	25.53	0.81
A10	71100239	21.33	1.36	27.08	2.44
A11	64444211	19.18	0.23	21.37	1.43
A12	29180006	20.52	2.58	29.01	0.71
A13	213400977	11.75	0.73	14.20	1.02
A14	226360542	17.98	0.99	20.69	1.03
A15	238633039	16.87	1.33	20.07	0.71
A16	37325677	22.37	0.00	23.59	1.21
A17	19623149	54.83	0.00	54.83	0.00
A18	60282102	22.42	2.00	27.06	2.25
A19	41044876	20.33	1.23	28.23	4.46
A20	14710475	35.25	0.00	35.25	0.00
Average		20.93	1.19	23.58	1.12

Table 3. Results for the Set B of instances

Instance	LB	BRKGA		EA	
		gap (%)	cv (%)	gap (%)	cv (%)
B1	77110392	10.12	1.00	9.11	0.24
B2	315354085	18.47	1.13	21.04	0.66
B3	349989487	15.36	0.72	16.79	0.62
B4	148205615	15.98	0.23	18.17	0.74
B5	319711555	37.81	0.00	37.81	0.00
B6	192874073	15.69	1.45	18.89	0.87
B7	187746291	11.82	1.41	17.17	1.03
B8	339397811	13.44	0.65	16.68	0.68
B9	293827643	12.06	0.61	14.35	0.82
B10	345904837	14.32	1.23	18.47	0.99
B11	336052870	14.34	0.84	18.81	0.53
B12	259876763	18.19	0.80	22.51	0.54
B13	484072875	17.81	0.94	22.36	0.66
B14	176124110	17.05	1.32	21.33	0.88
B15	432558079	17.07	0.71	20.20	0.91
Average		12.48	0.86	14.68	0.67

Table 4. Results for the Set X of instances

Instance	LB	BRKGA		EA	
		gap (%)	cv (%)	gap (%)	cv (%)
X1	244983403	17.38	0.76	17.93	0.74
X2	147062803	12.08	1.22	17.81	0.81
X3	156830774	15.93	1.16	19.03	1.33
X4	241257058	14.97	1.12	17.78	0.51
X5	78796003	19.99	1.05	22.86	0.59
X6	248147317	15.59	0.68	19.10	0.98
X7	369443070	15.72	0.88	18.85	0.77
X8	134427339	47.35	1.57	47.62	0.00
X9	361189695	15.49	1.06	19.53	1.09
X10	333756718	15.16	0.57	19.14	0.87
X11	223408308	19.19	1.12	23.14	1.26
X12	242393345	16.76	1.44	21.97	1.08
X13	259467828	13.91	0.78	19.41	0.98
X14	158409238	19.33	1.00	22.73	0.85
X15	242691036	19.02	1.71	24.73	1.27
Average		13.89	1.07	16.58	0.87

third and fourth columns present the results for the BRKGA. The third column presents the relative optimality gap of the BRKGA's fitness computed against the lower bound displayed in the second column, while the fourth column gives the coefficient of variation (cv) of the algorithm's results. The same information is given for the EA in the fifth and sixth columns, respectively. The last line of each table presents the average relative optimality gap and the average coefficient of variation for each heuristic.

One can see from these tables that the relative optimality gap of BRKGA was smaller or equal than that of EA for all evaluated instances. BRKGA obtained an average relative optimality gap of 20.93%, 12.48%, and 13.89% for the sets A, B, and X of instances, respectively, while that of EA was 23.58%, 14.68%, and 16.58%. Therefore, it can be concluded that BRKGA obtained better results than EA when solving the proposed instances. However, one can observe that EA is a more stable method than BRKGA, as its average coefficient of variation was smaller than that of BRKGA for all sets of evaluated instances.

7   Concluding Remarks

In this work, we tackled the Two-dimensional Three-staged Cutting Stock Problem with Stack Constraints (2DCSP-SC). We extended the Evolutionary Algorithm (EA) described in [21] to address the 2DCSP-SC. Furthermore, we proposed a Biased Random Key Genetic Algorithm (BRKGA). Both algorithms use the Finite First Fit (FFF) heuristic as a decoder. Computational experiments, performed on three sets of realistic instances, show that BRKGA found solutions with smaller optimally gaps in all but one of the instances tested.

Future works may explore exact methods, such as branch-and-bound algorithms, to improve the lower bounds proposed in this paper. Alternatively, other heuristic methods that do not rely in Genetic Algorithms could be devised for the problem, such as heuristics based on local search.

References

1. Parreño, F., Alonso, M., Alvarez-Valdes, R.: Solving a large cutting problem in the glass manufacturing industry. Eur. J. Oper. Res. **287**, 378–388 (2020)
2. Leao, A.A., Furlan, M.M., Toledo, F.M.: Decomposition methods for the lot-sizing and cutting-stock problems in paper industries. Appl. Math. Model. **48**, 250–268 (2017)
3. Galvez, J.L.A.P., Borenstein, D., da Silveira Farias, E.: Application of optimization for solving a sawing stock problem with a cant sawing pattern. Optim. Lett. **12**(8), 1755–1772 (2018)
4. Karelahti, J.: Solving the cutting stock problem in the steel industry. Helsinki University of Technology, pp. 1–39 (2002)
5. Israni, S., Sanders, J.: Two-dimensional cutting stock problem research: a review and a new rectangular layout algorithm. J. Manuf. Syst. **1**(2), 169–182 (1982)
6. Andrade, R., Birgin, E.G., Morabito, R.: Two-stage two-dimensional guillotine cutting stock problems with usable leftover. Int. Trans. Oper. Res. **23**(1–2), 121–145 (2016)

7. Aryanezhad, M., Hashemi, N.F., Makui, A., Javanshir, H.: A simple approach to the two-dimensional guillotine cutting stock problem. J. Ind. Eng. Int. **8**, 21 (2012)
8. Birgin, E.G., Romão, O.C., Ronconi, D.P.: The multiperiod two-dimensional non-guillotine cutting stock problem with usable leftovers. Int. Trans. Oper. Res. **27**(3), 1392–1418 (2020)
9. Clautiaux, F., Sadykov, R., Vanderbeck, F., Viaud, Q.: Pattern-based diving heuristics for a two-dimensional guillotine cutting-stock problem with leftovers. EURO J. Comput. Optim. **7**(3), 265–297 (2019)
10. Furini, F., Malaguti, E., Thomopulos, D.: Modeling two-dimensional guillotine cutting problems via integer programming. INFORMS J. Comput. **28**, 736–751 (2016)
11. Leung, T., Yung, C., Troutt, M.D.: Applications of genetic search and simulated annealing to the two-dimensional non-guillotine cutting stock problem. Comput. Ind. Eng. **40**(3), 201–214 (2001)
12. Alvarez-Valdes, R., Martí, R., Tamarit, J.M., Parajón, A.: Grasp and path relinking for the two-dimensional two-stage cutting-stock problem. INFORMS J. Comput. **19**(2), 261–272 (2007)
13. Hifi, M., Roucairol, C.: Approximate and exact algorithms for constrained (un) weighted two-dimensional two-staged cutting stock problems. J. Comb. Optim. **5**(4), 465–494 (2001)
14. Yanasse, H.H., Morabito, R.: Linear models for 1-group two-dimensional guillotine cutting problems. Int. J. Prod. Res. **44**(17), 3471–3491 (2006)
15. Lydia, T., Quentin, V.: Challenge ROADEF - euro 2018 cutting optimization problem description (2018). https://www.roadef.org/challenge/2018/files/Challenge_ROADEF_EURO_SG_Description.pdf
16. Cui, Y., Song, X., Chen, Y., Cui, Y.: New model and heuristic solution approach for one-dimensional cutting stock problem with usable leftovers. J. Oper. Res. Soc. **68**(3), 269–280 (2017)
17. Chen, Q., Chen, Y., Cui, Y., Lu, X., Li, L.: A heuristic for the 3-staged 2D cutting stock problem with usable leftover. In: 2015 International Conference on Electrical, Automation and Mechanical Engineering. Atlantis Press, Phuket, pp. 776–779 (2015)
18. Cherri, A.C., Arenales, M.N., Yanasse, H.H.: The one-dimensional cutting stock problem with usable leftover-a heuristic approach. Eur. J. Oper. Res. **196**(3), 897–908 (2009)
19. Lai, K.K., Chan, J.: Developing a simulated annealing algorithm for the cutting stock problem. Comput. Ind. Eng. **32**, 115–127 (1997)
20. Silva, E., Alvelos, F., De Carvalho, J.V.: An integer programming model for two-and three-stage two-dimensional cutting stock problems. Eur. J. Oper. Res. **205**(3), 699–708 (2010)
21. Puchinger, J., Raidl, G.R., Koller, G.: Solving a real-world glass cutting problem. In: Gottlieb, J., Raidl, G.R. (eds.) EvoCOP 2004. LNCS, vol. 3004, pp. 165–176. Springer, Heidelberg (2004). https://doi.org/10.1007/978-3-540-24652-7_17
22. Suliman, S.: A sequential heuristic procedure for the two-dimensional cutting-stock problem. Int. J. Prod. Econ. **99**(1–2), 177–185 (2006)
23. Jin, M., Ge, P., Ren, P.: A new heuristic algorithm for two-dimensional defective stock guillotine cutting stock problem with multiple stock sizes. Tehnicki Vjesnik **22**, 1107–1116 (2015)
24. Vassiliadis, V.S.: Two-dimensional stock cutting and rectangle packing: binary tree model representation for local search optimization methods. J. Food Eng. **70**(3), 257–268 (2005)

25. Dusberger, F., Raidl, G.R.: Solving the 3-staged 2-dimensional cutting stock problem by dynamic programming and variable neighborhood search. Electron. Notes Discrete Math. **47**, 133–140 (2015)
26. Dusberger, F., Raidl, G.R.: A variable neighborhood search using very large neighborhood structures for the 3-staged 2-dimensional cutting stock problem. In: Blesa, M.J., Blum, C., Voß, S. (eds.) Hybrid Metaheuristics, pp. 85–99. Springer, Cham (2014)
27. Hifi, M., M'Hallah, R., Saadi, T.: Algorithms for the constrained two-staged two-dimensional cutting problem. INFORMS J. Comput. **20**(2), 212–221 (2008)
28. Davis, L.: Handbook of Genetic Algorithms, ser. VNR Computer Library VNR Computer Library. Van Nostrand Reinhold, New York (1991). https://books.google.com.br/books?id=Kl7vAAAAMAAJ
29. Bean, J.C.: Genetic algorithms and random keys for sequencing and optimization. ORSA J. Comput. **2**, 154–160 (1994)
30. Gonçalves, J.F., Resende, M.G.C.: Biased random-key genetic algorithms for combinatorial optimization. J. Heuristics **17**, 487–525 (2011)
31. Spears, W., deJong, K.: On the virtues of parameterized uniform crossover. In: Belew, R., Booker, L. (eds.) Proceedings of the Fourth International Conference on Genetic Algorithms. Morgan Kaufman, San Mateo, pp. 230–236 1991)
32. Matsumoto, M., Nishimura, T.: Mersenne twister: a 623-dimensionally equidistributed uniform pseudo-random number generator. ACM Trans. Model. Comput. Simul. (TOMACS) **8**(1), 3–30 (1998)

A Genetic Algorithm for a Capacitated Lot-Sizing Problem with Lost Sales, Overtimes and Safety Stock Constraints

Benoît Le Badezet[1], François Larroche[1,2(✉)], Odile Bellenguez[1],
and Guillaume Massonnet[1]

[1] IMT Atlantique, LS2N, La Chantrerie, 4 Rue Alfred Kastler, 44307 Nantes, France
benoit.le-badezet@etu.univ-nantes.fr, {francois.larroche,
odile.bellenguez,guillaume.massonnet}@imt-atlantique.fr
[2] VIF, 10 Rue de Bretagne, 44240 La Chapelle-sur-Erdre, France

Abstract. This paper deals with a complex production planning problem with lost sales, overtimes, safety stock and sequence dependent setup times on parallel and unrelated machines. The main challenge of this work is to propose a solution approach to obtain a good feasible plan in a short execution time (around 2 min) for large industrial instances. We develop a genetic algorithm that combines several operations already defined in the literature to solve the problem. Preliminary numerical results obtained with our algorithm are presented and compared to a straightforward MIP resolution. The method appears to be an appealing alternative on large instances when the computational time is limited.

1 Introduction

The problem presented in this paper is related to practical cases encountered in the food industry for production planning. In this context, manufacturers can generally use several production lines, each able to make several types of items. This complexity usually leads to problems that are too large to be solved optimally by off-the-shelf solvers. In addition, the models we consider in this paper also combines constraints from the lot-sizing and the scheduling literature, by assuming that the setup times between different types of items depends on the production sequence. This further limit the applicability of standard methods in practice, when the planners need to obtain "good" feasible solutions in reasonable time to test several machine configurations or shifts assignments and obtain quick insights to support their decisions.

This problem extends the field of lot-sizing, which has been extensively studied since the work of Wagner and Whitin [1]. Motivated by the physical constraints found in practical applications, the finite production capacity version of the problem (CLSP) has received a lot of attention, see [2] and [3] for a review of extensions and solution approaches. The problem we consider is an extension of the industrial problem with lost sales and shortage costs presented in [4], for which the authors introduce new classes of valid inequalities. The safety stock

B. Dorronsoro et al. (Eds.): META 2021, CCIS 1541, pp. 170–181, 2022.
https://doi.org/10.1007/978-3-030-94216-8_13

is seldom considered in the deterministic production and inventory literature. [5] define the safety stock as a lower bound on the number of units that must be held in the inventory at each period when [6] choose to penalize the missing units from the safety stock. The latest version is studied here. Versions of the problem with parallel machines and sequence dependent setups are less common in the literature. [7] develop new heuristics on a parallel machines problems. [8] present an industrial problem in which setup times depend on the sequence of production and propose a solution procedure based on subtour elimination and patching. [9] use a small bucket formulation to compute the sequence of production. [10] also present an extensive review of this extension and compare the efficiency of several methods to solve it. The possibility to exceed the production capacity is not common in the literature, see [11] for an overtime extension of a capacitated lot-sizing problem.

In terms of metaheuristics, various researches have been done on the previously detailed extensions of our problem. [12] propose a Genetic Algorithm (GA) to tackle a multi-items CLSP and on multiple production lines, using various crossovers and mutation. The authors also use a new operator called "siblings" that consists in a local search using a ranking system on the neighbours. [13] propose a Tabu-Search (TS) to solve the same problem. [14] and [15] propose hybridized GA to solve the CLSP with an overtime constraints. The hybridization introduces elements of Tabu-search and Simulated Annealing into the GA in order to improve the efficiency of the algorithm. On top of that, they also use multi-population on different version of the algorithm to tackle their instances. On the single-machine CLSP with sequence dependent setup times, [16] and [17] propose a Threshold Accepting whereas [18] develop a Tabu-Search and [19] propose a GA. To the best of our knowledge however, none of the previous problems incorporate a target-stock constraint similar to our case.

In the following, we denote CLSSD-PM the *multi-item capacitated lot-sizing problem with lost sales, safety stock, overtimes, and sequence dependent setups on parallel machines*. A previous work in [20] focus on a part of this problem without safety stock. To the best of our knowledge, this whole problem has never been studied in the literature before.

2   Problem Definition

The CLSSD-PM is a extensive version of the capacitated lot-sizing problem which is proven to be NP-hard [21]. The goal is to plan the production of N different items, over T time periods and on M parallel unrelated production lines. There is a demand d_t^i for each item $i \in \{1, ..., N\}$ in each period $t \in \{1, ..., T\}$ that must be satisfied if units of i are available in stock. When that is not the case, the demand can be (partially or totally) lost, incurring a per-unit lost sales cost l_t^i. Any production of item i in period t on line $m \in \{1, ..., M\}$ is an integral number of batches, i.e. a multiple of a fixed quantity Q^i of units. The production of one such batch incurs a cost p_{mt}^i and requires a production time τ_m^i. In addition, the production of items of type k immediately after items of type $i \neq k$ during a given period on machine m induces a setup time γ_m^{ik}.

Each line m at each period t has a (planned) time capacity of C_{mt}, but production overtimes are allowed up to a maximum total production time $\bar{C}_{mt}$. When production occurs during the planned capacity, the corresponding cost of line usage is c_{mt} per unit of time, but this cost increases to $c_{mt} + \bar{c}_{mt}$ when the production needs overtime, i.e. for any usage that exceeds C_{mt}.

We model item storage by the mean of a target stock S_{it} for each item i in each period t. Any unit of stock of i in period t that exceeds S_{it} induces an excess storage cost of h_{it}^{+}, while missing inventory to reach the target stock incurs a per-unit penalty equal h_{it}^{-}.

We also make the following hypothesis on our problem:

- Demand and inventory are satisfied and consumed following a FIFO rule. This implies that it is impossible to choose to loose some demand of an item that is held in stock.
- Setup times between items follow the triangle inequality rule.
- At the beginning of each period, each line is in a neutral state, and the setup time to start the production of the first item in any period is null.

The objective of our problem is to minimize the total cost of the production planning (line usage, production, storage and lost sales combined). For conciseness reasons, we do not present the MILP formulation here and instead refer the interested readers to the [22].

3   Genetic Algorithm

We now develop a genetic algorithm to address the CLSSD-PM. We start by introducing the general structure of the procedure, before presenting in mode details the chromosome representation, crossover and mutation operators.

3.1   Genetic Algorithm Pseudo-Code

We use a generational genetic algorithm (GA), which creates successive generations of a population of individuals, by using specific operators inspired by nature and called crossovers and mutations. We keep some overlapping between consecutive generations, i.e. some of the best elements obtained in the current population are retained for the next generation, to keep the most interesting information of what has been done in previous iterations.

To avoid being in a local optimum for too long, the algorithm sometimes performs a reset that re-generates randomly a large portion of the current population. This operation is done only after a long period without improvement of the best known solution. A pseudo-code of the procedure is presented in Algorithm 1.

Algorithm 1: *Genetic Algorithm*

---

1 curGen ← GeneratePopulation() ;
2 s^* ← bestIndividual ;
3 **while** *stopping criterion not met* **do**
4   nextGen ← overlap(curGen);
5   **while** $|nextGen| < maxPopSize$ **do**
6   **if** *condCrossover* **then**
7   parent1, parent2 ← selectionCross(curGen);
8   nextGen.add(crossover(parent1, parent2));
9   **end**
10   **if** *condMutation* **then**
11   mutated ← selectionMutation(curGen);
12   nextGen.add(mutation(mutated));
13   **end**
14   **end**
15   **if** *condReset* **then**
16   reset();
17   **end**
18   **if** *cost(nextGen.bestIndividual) < cost(s*)* **then**
19 s^* ← nextGen.bestIndividual ;
20   **end**
21   curGen ← nextGen ;
22 **end**
23 **return** s^*

3.2 Chromosome Representation

The problem requires two type of decisions: The first one assigns the production of items to periods and machines, while the second one aims at designing the production sequences. As a consequence, we propose the following independent variables that serve as chromosomes:

- x_{mt}: Set of tuples ⟨item; quantity⟩ produced on m during t.
- w_{mt}: Contains the ordered sequence of production on m during t.

Other necessary information to represent a solution are deduced from these two variables, using the dependent variables below:

- $cost$: Total cost of the solution.
- u_{mt}: Time usage of line m in period t.
- $prod_t^i$: Number of batches of i produced in period t.
- $stock_t^i$: Stock of item i available at the end of period t.
- L_t^i: Number of lost sales for item i in period t.

174   B. Le Badezet et al.

To ensure diversity within the population, we start a completely random chromosome generation. For each line in each period we draw randomly a subset of items and affect to each of them a random production quantity. The sequence is determined as the items are drawn. Since the goal is to minimize the objective function of our problem, we keep a fitness parameter $fitness = \frac{1}{cost(solution)}$ updated to ensure that the gaps between the costs of different solutions are proportional. Finally, the selection is made based on a roulette wheel mechanism, applied to the fitness of the population.

3.3 Crossover

In order to explore a large variety of solutions, we apply several crossovers from one generation to the next, in a similar fashion as the GA presented in [12]. In our case, we have three different crossover:

- On periods.
- On items.
- On sequences.

Crossover on Periods. This crossover is heavily inspired by [12]. It basically consists of a two-point crossover applied on the periods of the solutions. The concept is to choose randomly a subset of periods and exchange all the production quantities of the two parents in the selected periods. Table 1 illustrates a crossover on periods 3 and 4.

Table 1. Illustration of the period crossover

Crossover on Items. This crossover is inspired by [12] For a given machine m and a given time period t, this crossover iterates following the item information stored in the chromosomes x_{mt} of both parents. For each m and t, we consider the union the items produced by the two parents and draw for each of them a random boolean. If we draw 0 then the first child takes the first parent's production, and the second child takes the second parent's. Otherwise the first child takes the second parent's production, and the second child takes the first parent's. Table 2 shows a practical example of this crossover for a given period and line.

Table 2. Example of item crossover for a given period and line

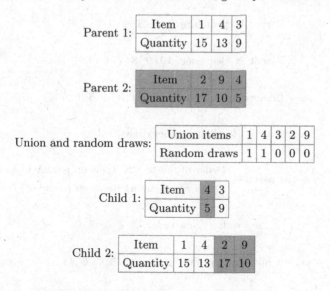

Parent 1:

Item	1	4	3
Quantity	15	13	9

Parent 2:

Item	2	9	4
Quantity	17	10	5

Union and random draws:

Union items	1	4	3	2	9
Random draws	1	1	0	0	0

Child 1:

Item	4	3
Quantity	5	9

Child 2:

Item	1	4	2	9
Quantity	15	13	17	10

Crossover on the Sequences. This crossover is inspired by [19] This crossover enables us to change the sequence of production. For each line and in each period, we form the set containing the common items from the two parent solutions. We then create a new sequence in the following manner:

1. Draw a random integer X between 1 and the number of common items
2. Order the X first item as they are in the sequence of the first parent. The remaining items follow the same order they have in the sequence of the second parent.
3. Create the sequences of the children using the parents sequences in which the common items are reordered.

Table 3 shows a practical example of this crossover for a given period and line.

3.4   Mutation

We consider a mutation that swaps the positions of two randomly selected items in the sequence of production, as represented in Table 4. As we also do not want to alter the totality of the individual we will add a parameter to describe the amount of information that will be altered in a mutated individual.

Table 3. Example of sequence crossover for a given period and line

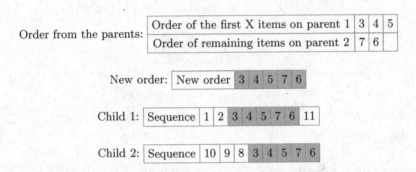

Parent 1: | Sequence | 1 | 2 | 3 | 4 | 5 | 6 | 7 | 11 |

Parent 2: | Sequence | 10 | 9 | 8 | 7 | 6 | 5 | 4 | 3 |

Intersection: | Common items | 3 | 4 | 5 | 6 | 7 |

Random number draw X: 3

Order from the parents:

| Order of the first X items on parent 1 | 3 | 4 | 5 |
| Order of remaining items on parent 2 | 7 | 6 | |

New order: | New order | 3 | 4 | 5 | 7 | 6 |

Child 1: | Sequence | 1 | 2 | 3 | 4 | 5 | 7 | 6 | 11 |

Child 2: | Sequence | 10 | 9 | 8 | 3 | 4 | 5 | 7 | 6 |

Table 4. Example of a mutated sequence on items 2 and 6

| Sequence before mutation | 1 | 2 | 3 | 4 | 5 | 6 |
| Sequence after mutation | 1 | 6 | 3 | 4 | 5 | 2 |

3.5   Repair

Note that such movements may result in infeasible solution since some line usage may exceed its maximum capacity. When this situation arise, we repair them by removing the production of one or more items until we don't exceed the hard capacity anymore and then replace it if possible on previous periods. In order to have a minimum impact on the quality of the solution, we chose to remove the item having the highest ratio $\frac{prod_t^i}{demand_t^i}$ so that we can avoid most of the lost sales. The quantity the remove in order to make the period feasible, is stored and will be spread on the previous periods where the item was already in production.

4   Experimentation Results

The instances that we use for our numerical experiments are derived from practical applications defined by VIF, a software company specialised in solutions for the food industry.

4.1   Parameters

Our algorithm is tuned through 9 parameters that have been tested to choose the best possible values.

- Size of the population: 200 individuals.
- Number of generations: 15000 generations, limited to 2 min of execution.
- Percentage of overlapping population between generations: 10%.
- Percentage of rested population: 50%.
- Number of non-improving iterations needed to reset: 200.
- Crossover ratio: 90%.
- Mutation ratio: 10%.
- Percentage of information of an individual that will be mutated: 20%.
- Ratio between the different crossovers: 60% period crossover, 20% item crossover, and 20% sequence crossover.

4.2   Experimentation

Implementation and tests of the algorithms have been done in Java. Tests have been realised on a personal computer with the following characteristics:

OS: Ubuntu 18.04.4 LTS
Processor: Intel i5-7600K @ 4.200 GHz × 4
GPU: NVIDIA GeForce GTX 1070
RAM: 16 Gb
Type: 64-bit

We tested our GA on 168 instances that combine the following parameters: Number of items $\in \{20, 30, 40, 50, 75, 100, 125\}$, number of lines $\in \{1, 2, 4, 6\}$ and number of periods $\in \{15, 30\}$. The lower bound and upper bound considered are based on the results computed by CPLEX in 4 h using a MIP formulation of the problem. We compare our results with the best lower bound (LB) obtained by CPLEX using settings presented in [22] and compute the gap achieved by our procedure with the following formula:

$$Gap = \frac{GA.cost - LB}{LB} \times 100$$

4.3   Results

We tested the GA presented in this paper with a maximum computational time of 2 min and compared the solutions obtained with the ones found by CPLEX in 4 h. Note that the latter are used as a baseline and do not represent a viable option for practitioners to do its large computational time. In fact except for the smallest instances, CPLEX rarely even finds a feasible solution within 2 min, which already gives the GA an edge in the specific application that is targeted. In

Table 5. Comparison of gaps obtained by CPLEX (4 h) and our GA (2 min) by group of same size instances

Instances	CPLEX Gap(%)			GA Gap(%)		
Items-lines-periods	Min	Max	Mean	Min	Max	Mean
20-1-15	0.1	4.1	**1.7**	381.9	625.8	**534.0**
20-1-30	1.8	19.9	**8.9**	492.1	836.4	**731.4**
20-2-15	0.1	10.0	**2.5**	258.0	520.3	**374.2**
20-2-30	2.4	13.2	**7.0**	527.8	1108.7	**707.8**
30-1-15	1.5	12.0	**5.0**	602.4	1 203.1	**907.3**
30-1-30	0.8	1 163.3	**277.1**	1 000.8	1 911.9	**1 243.8**
30-2-15	1.8	18.8	**9.5**	513.2	1 122.8	**792.2**
30-2-30	7.4	544.4	**190.9**	379.9	1 298.2	**975.5**
40-1-15	7.6	75.1	**40.2**	685.7	1 471.5	**1 094.8**
40-1-30	533.8	2 853.7	**1 198.4**	685.0	2 245.4	**1 352.4**
40-2-15	6.0	789.9	**145.7**	391.5	1 553.2	**1 025.3**
40-2-30	322.2	7 127.8	**3 177.8**	640.1	1 768.2	**1 383.7**
50-1-15	83.0	3 573.4	**842.4**	968.8	1 778.6	**1 411.2**
50-1-30	292.0	14 145.4	**4 063.2**	998.6	3 452.6	**2 033.9**
50-2-15	4.9	1 305.0	**735.0**	1 014.6	1 598.5	**1 351.5**
50-2-30	530.9	4 277.8	**2 543.6**	1 681.4	2 312.7	**1 914.2**
75-2-15	774.5	2 811.1	**1 713.8**	2 070.9	3 459.9	**2 891.2**
75-2-30	432.6	6 829.5	**2 879.3**	2 139.9	4518.3	**3 328.3**
75-4-15	464.6	2 134.6	**1 744.8**	1 883.3	3 079.3	**2 582.1**
75-4-30	3 141.6	7 866.4	**4 537.1**	1 912.0	4 281.5	**3 387.3**
100-2-15	1 163.1	4 824.4	**2 494.2**	2 055.4	5 426.2	**3 864.5**
100-2-30	4 406.2	8 608.9	**6 371.8**	3 921.6	5 789.1	**4 449.2**
100-4-15	742.3	5 212.6	**2 569.0**	2 787.8	4 832.3	**4 034.2**
100-4-30	3 243.8	6 843.9	**5 090.9**	4 273.1	6 017.9	**5 241.9**
125-4-15	3 126.3	11 547.6	**5 756.7**	3 945.9	6 101.4	**4 935.5**
125-4-30	4 960.9	19 640.6	**8 306.3**	5 348.4	7 110.8	**6 402.2**
125-6-15	2 437.1	6 862.7	**4 358.6**	3 169.0	6 314.2	**5 301.0**
125-6-30	5 021.4	17 563.0	**7 579.5**	5 423.7	7 130.9	**6 295.3**

addition we observe that in 45 out of the 168 tested instances, our GA obtains a better solution in 2 min than the one obtained by CPLEX in 4 h. The distribution of these 45 instances is as follows:

- 0 case for 20 items.
- 1 case for 30 items (0 for 15 periods, 1 for 30 periods).
- 7 cases for 40 items (0 for 15 periods, 7 for 30 periods).
- 10 cases for 50 items (2 for 15 periods, 8 for 30 periods).

- 9 cases for 75 items (3 for 15 periods, 6 for 30 periods).
- 11 cases for 100 items (7 for 15 periods, 4 for 30 periods).
- 7 cases for 125 items (5 for 15 periods, 2 for 30 periods).

The Table 5 compares the gaps obtained by CPLEX and our GA on groups of 6 instances of same size. For each group we retain the minimal gap obtained, the maximal gap and the mean gap for all 6 instances.

This table also shows clearly the great differences that can appear between solutions found by CPLEX on 2 instances of same size (example for instance of size 50-1-30 where we have a minimal gap of 292% and a maximal gap of 14 145%) whereas our GA shows closer values (min: 999%, max: 3 453%). In general, the consistency of the results obtained by the GA is better across instances of the same size: In particular it appears that the solutions from CPLEX seem more sensitive to the number of periods that our procedure. Even if the results obtained by our GA are far behind the ones obtained by the MIP for the smallest instances, they become competitive on larger ones. For the largest instances, our heuristic consistently outperforms in 2 min the feasible solution computed by CPLEX in 4 h.

These results clearly demonstrate the tendency of metaheuristics, in this case a genetic algorithm, to deal quickly with complex problems, and their usefulness in practice to tackle large industrial instances compared to MIP formulations and commercial solvers. Finally, note that the two approaches can also be used in combination, where the solution find by the GA can serve as a first feasible solution for the MIP solver, in an attempt to speed up the its convergence towards an optimal solution.

5   Conclusion

In this work, we apply the well-known genetic algorithm paradigm to develop a dedicated algorithm that is able to run quickly on large industrial instances of a complex practical production planning problem. The main contributions of this study can be partitioned in two broad categories. First, the heuristic developed is the first one that takes into account several industrial extensions of classical lot-sizing problems, such as the combination of multiple unrelated machines and sequence-dependent setup times. Second, it provides a viable alternative to commercial solvers to deal with large industrial instances that displays a robust behavior with respect to the size of the problem considered. Note that the solution obtained using our procedure may serve as a warm start for an exact method.

While the first results obtained show that such metaheuristics are a viable alternative on large instances, additional work is necessary to improve the overall performances. In particular, the method would become a lot more reliable if the solutions on small instances were comparable to the ones computed by commercial solvers. Local search methods or more advanced concepts such as hybridization or multi-population could help reduce the gap in such cases. We could also seek to find dominance properties to reduce the search space and speed up the resolution.

Another research direction to achieve this goal is to apply the procedure to a simpler problem that approximates the original one. In a recent paper [22], we developed a procedure that computes clusters of items with small switching times, which enables the algorithm to primarily focus on positioning clusters in the production sequence rather than items. This approximation greatly reduces the size of the original problem and was proven successful when used in combination with classical heuristics from the lot-sizing literature. It is likely that the GA presented in this paper would also benefit from this reduction of the problem size to converge faster to good quality solutions.

References

1. Wagner, H.M., Whitin, T.M.: Dynamic version of the economic lot size model. Manage. Sci. **5**, 89–96 (1958)
2. Quadt, D., Kuhn, H.: Capacitated lot-sizing with extensions: a review. 4OR **6**, 61–83 (2008). https://doi.org/10.1007/s10288-007-0057-1
3. Karimi, B., Fatemi Ghomi, S., Wilson, J.: The capacitated lot sizing problem: a review of models and algorithms. Omega **31**, 365–378 (2003)
4. Absi, N., Kedad-Sidhoum, S.: The multi-item capacitated lot-sizing problem with setup times and shortage costs. Eur. J. Oper. Res. **185**, 1351–1374 (2008)
5. Loparic, M., Pochet, Y., Wolsey, L.A.: The uncapacitated lot-sizing problem with sales and safety stocks. Math. Program. **89**, 487–504 (2001)
6. Absi, N., Kedad-Sidhoum, S.: The multi-item capacitated lot-sizing problem with safety stocks and demand shortage costs. Comput. Oper. Res. **36**, 2926–2936 (2009)
7. Beraldi, P., Ghiani, G., Grieco, A., Guerriero, E.: Rolling-horizon and fix-and-relax heuristics for the parallel machine lot-sizing and scheduling problem with sequence-dependent set-up costs. Comput. Oper. Res. **35**, 3644–3656 (2008)
8. Clark, A.R., Morabito, R., Toso, E.A.V.: Production setup-sequencing and lot-sizing at an animal nutrition plant through ATSP subtour elimination and patching. J. Sched. **13**, 111–121 (2010). https://doi.org/10.1007/s10951-009-0135-7
9. Gicquel, C., Minoux, M., Dallery, Y., Blondeau, J.M.: A tight MIP formulation for the discrete lot-sizing and scheduling problem with parallel resources. In: 2009 International Conference on Computers & Industrial Engineering, pp. 1–6. IEEE (2009)
10. Guimarães, L., Klabjan, D., Almada-Lobo, B.: Modeling lotsizing and scheduling problems with sequence dependent setups. Eur. J. Oper. Res. **239**, 644–662 (2014)
11. Özdamar, L., Bozyel, M.A.: The capacitated lot sizing problem with overtime decisions and setup times. IIE Trans. **32**, 1043–1057 (2000)
12. Hung, Y.F., Shih, C.C., Chen, C.P.: Evolutionary algorithms for production planning problems with setup decisions. J. Oper. Res. Soc. **50**, 857 (1999)
13. Hung, Y.F., Chen, C.P., Shih, C.C., Hung, M.H.: Using tabu search with ranking candidate list to solve production planning problems with setups. Comput. Ind. Eng. **45**, 615–634 (2003)
14. Özdamar, L., Birbil, Ş.İ.: Hybrid heuristics for the capacitated lot sizing and loading problem with setup times and overtime decisions. Eur. J. Oper. Res. **110**, 525–547 (1998)

15. Özdamar, L., Bilbil, Ş.İ., Portmann, M.C.: Technical note: New results for the capacitated lot sizing problem with overtime decisions and setup times. Prod. Plann. Control **13**, 2–10 (2002)
16. Fleischmann, B., Meyr, H.: The general lotsizing and scheduling problem. Oper.-Res.-Spektrum **19**(1), 11–21 (1997). https://doi.org/10.1007/BF01539800
17. Meyr, H.: Simultaneous lotsizing and scheduling by combining local search with dual reoptimization. Eur. J. Oper. Res. **120**, 311–326 (2000)
18. Laguna, M.: A heuristic for production scheduling and inventory control in the presence of sequence-dependent setup times. IIE Trans. **31**, 125–134 (1999)
19. Sikora, R.: A genetic algorithm for integrating lot-sizing and sequencing in scheduling a capacitated flow line. Comput. Ind. Eng. **30**, 969–981 (1996)
20. Larroche, F., Bellenguez-Morineau, O., Massonnet, G.: Approche de résolution d'un problème industriel de lot-sizing avec réglages dépendant de la séquence. In: ROADEF 2020: 21ème Congrès Annuel de la Société Française de Recherche Opérationnelle et d'Aide á la Décision, Montpellier, France (2020)
21. Florian, M., Lenstra, J.K., Rinnooy Kan, A.H.G.: Deterministic production planning: algorithms and complexity. Manage. Sci. **26**, 669–679 (1980)
22. Larroche, F., Bellenguez, O., Massonnet, G.: Clustering-based solution approach for a capacitated lot-sizing problem on parallel machines with sequence-dependent setups*. Int. J. Prod. Res. **0**, 1–24 (2021)

GA and ILS for Optimizing the Size of NFA Models

Frédéric Lardeux[✉] and Eric Monfroy

Univ Angers, LERIA, SFR MATHSTIC, 49000 Angers, France
{frederic.lardeux,eric.monfroy}@univ-angers.fr

Abstract. Grammatical inference consists in learning a formal grammar (as a set of rewrite rules or a finite state machine). We are concerned with learning Nondeterministic Finite Automata (NFA) of a given size from samples of positive and negative words. NFA can naturally be modeled in SAT. The standard model [1] being enormous, we also try a model based on prefixes [2] which generates smaller instances. We also propose a new model based on suffixes and a hybrid model based on prefixes and suffixes. We then focus on optimizing the size of generated SAT instances issued from the hybrid models. We present two techniques to optimize this combination, one based on Iterated Local Search (ILS), the second one based on Genetic Algorithm (GA). Optimizing the combination significantly reduces the SAT instances and their solving time, but at the cost of longer generation time. We, therefore, study the balance between generation time and solving time thanks to some experimental comparisons, and we analyze our various model improvements.

Keywords: Constraint problem modeling · Grammar inference · SAT · Model reformulation · NFA inference

1 Introduction

Grammatical inference [3] (or grammar induction) is concerned with the study of algorithms for learning automata and grammars from some observations. The goal is thus to construct a representation that accounts for the characteristics of the observed objects. This research area plays a significant role in numerous applications, such as compiler design, bioinformatics, speech recognition, pattern recognition, machine learning, and others.

In this article, we focus on learning a finite automaton from samples of words $S = S^+ \cup S^-$, such that S^+ is a set of positive words that must be accepted by the automaton, and S^- is a set of negative words to be rejected by the automaton. Due to their determinism, deterministic finite automata (DFA) are generally faster than non deterministic automata (NFA). However, NFA are significantly smaller than DFA in terms of the number of states. Moreover, the space complexity of the SAT models representing the problem is generally due to the number of states. Thus, we focus here on NFA inference. An NFA is represented by a 5-tuple $(Q, \Sigma, \Delta, q_1, F)$ where Q is a finite set of states, the

© Springer Nature Switzerland AG 2022
B. Dorronsoro et al. (Eds.): META 2021, CCIS 1541, pp. 182–197, 2022.
https://doi.org/10.1007/978-3-030-94216-8_14

vocabulary Σ is a finite set of symbols, the transition function $\Delta : Q \times \Sigma \rightarrow \mathcal{P}(Q)$ associates a set of states to a given state and a given symbol, $q_1 \in Q$ is the initial state, and $F \subseteq Q$ is the set of final states.

The problem of inferring NFA has been undertaken with various approaches (see, e.g., [1]). Among them, we can cite ad-hoc algorithms such as *DeLeTe2* [4] that is based on state merging methods, or the technique of [5] that returns a collection of NFA. Some approaches use metaheuristics for computing NFA, such as hill-climbing [6] or genetic algorithm [7].

A convenient and declarative way of representing combinatorial problems is to model them as a Constraint Satisfaction Problem (CSP [8]) (see, e.g., [1] for an INLP model for inferring NFA, or [9] for a SAT (the propositional satisfiability problem [10]) model of the same problem). Parallel solvers have also been used for minimizing the inferred NFA size [2,11].

Orthogonally to the approaches cited above, we do not seek to improve a solver, but to generate a model of the problem that is easier to solve with a standard SAT solver. Our approach is similar to DFA inference with graph coloring [12], or NFA inference with complex data structures [9]. Modeling thus consists in translating a problem into a CSP made of decision variables and constraints over these variables. As a reference for comparisons, we start with the basic SAT model of [9]. The model, together with a sample of positive and negative words, lead to a SAT instance to be solved by a classic SAT solver that we use as a black box. However, SAT instances are gigantic, e.g., our base model space complexity is in the order of $\mathcal{O}(k^{|\omega_+|})$ variables, and in $\mathcal{O}(|\omega_+|.k^{|\omega_+|})$ clauses, where k is the number of states of the NFA, and ω_+ is the size of the longest positive word of the sample. The second model, PM, is based on intermediate variables for each prefix [2] which enables to compute only once parts of paths that are shared by several words. We propose a third model, SP, based on intermediate variables for suffixes. Although the two models could seem similar, their order of size is totally different. Indeed, PM is in $\mathcal{O}(k^2)$ while SM is in $\mathcal{O}(k^3)$. We then propose hybrid models consisting in splitting words into a prefix and a suffix. Modeling the beginning of the word is made with PM while the suffix is modeled by SM. The challenge is then to determine where to split words to optimize the size of the generated SAT instances. To this end, we propose two approaches, one based on iterated local search (ILS), the second one on genetic algorithm (GA). Both permit to generate smaller SAT instances, much smaller than with the DM model and even the PM model. However, with GA, the generation time is too long and erases the gain in solving with the Glucose SAT solver [13]. But the hybrid instances optimized with the ILS are smaller, and the generation time added to the solving time is faster than with PM. Compared to [9], which is the closest work on NFA inferring, we always obtain significantly smaller instances and solving time.

This paper is organized as follows. In Sect. 2 we present the direct model, the prefix model, and we propose the suffix model. We then combine suffix and prefix model to propose the new hybrid models (Sect. 3). Hybrid models are optimized with iterated local search (Subsect. 3.2), and with genetic algorithm in Subsect. 3.3. We then compare experimentally our models in Sect. 4 before concluding in Sect. 5.

2   SAT Models

Given an alphabet $\Sigma = \{s_1, \ldots, s_n\}$ of n symbols, a training sample $S = S^+ \cup S^-$, where S^+ (respectively S^-) is a set of *positive words* (respectively *negative words*) from Σ^*, and an integer k, **the NFA inference problem** consists in building a NFA with k states which validates words of S^+, and rejects words of S^-. Note that the satisfaction problem we consider in this paper can be extended to an optimization problem minimizing k [2].

Let us introduce some notations. Let $A = (Q, \Sigma, q_1, F)$ be a NFA with: $Q = \{q_1, \ldots, q_k\}$ a set of k states, Σ a finite alphabet, q_1 the initial state, and F the set of final states. The empty word is noted λ. We denote by K the set of integers $\{1, \ldots, k\}$.

We consider the following variables:

- k the size of the NFA we want to learn,
- a set of k Boolean variables $F = \{f_1, \ldots, f_k\}$ determining whether states q_1 to q_k are final or not,
- and $\Delta = \{\delta_{s, \overrightarrow{q_i q_j}} | s \in \Sigma \text{ and } i, j \in K\}$ a set of $n.k^2$ Boolean variables defining the existence or not of the transition from state q_i to state q_j with the symbol s, for each q_i, q_j, and s.

The path $i_1, i_2, \ldots, i_{n+1}$ for $w = w_1 \ldots w_n$ exists if and only if $d = \delta_{w_1, \overrightarrow{q_{i_1} q_{i_2}}} \wedge \ldots \wedge \delta_{w_n, \overrightarrow{q_{i_n} q_{i_{n+1}}}}$ is true. We say that the conjunction d is a c_path, and $D_{w, \overrightarrow{q_i q_j}}$ is the set of all c_paths for the word w between states q_i and q_j.

2.1   Direct Model

This simple model has been presented in [9]. It is based on 3 sets of equations:

1. If the empty word is in S^+ or S^-, we can fix whether the first state is final or not:

$$\text{if } \lambda \in S^+, \qquad f_1 \qquad\qquad (1)$$
$$\text{if } \lambda \in S^-, \qquad \neg f_1 \qquad\qquad (2)$$

2. For each word $w \in S^+$, there is at least a path from q_1 to a final state q_j:

$$\bigvee_{j \in K} \bigvee_{d \in D_{w, \overrightarrow{q_1 q_j}}} (d \wedge f_j) \qquad\qquad (3)$$

With the Tseitin transformations [14], we create one auxiliary variable for each combination of a word w, a state $j \in K$, and a c_path $d \in D_{w, \overrightarrow{q_1 q_j}}$: $aux_{w,j,d} \leftrightarrow d \wedge f_j$. Hence, we obtain a formula in CNF for each w:

$$\bigwedge_{j \in K} \bigwedge_{d \in D_{w, \overrightarrow{q_1 q_j}}} [(\neg aux_{w,j,d} \vee (d \wedge f_j))] \qquad\qquad (4)$$

Table 1. Clauses for DM_k

Number of cl.	Arity	Constraints
$|S^+|.(|\omega_+|+1).k^{|\omega_+|}$	2	(4)
$|S^+|.k^{|\omega_+|}$	$|\omega_+|+2$	(5)
$|S^+|$	$k^{|\omega_+|}$	(6)
$|S^-|.k^{|\omega_-|}$	$|\omega_-|+1$	(7)

Table 2. Variables for DM_k

Number of var	Reason
k	Final states F
$n.k^2$	Transitions δ
$|S^+|.k.^{|\omega_+|}$	Constraints (3)

$$\bigwedge_{j \in K} \bigwedge_{d \in D_{w,\overrightarrow{q_1 q_j}}} (aux_{w,j,d} \vee \neg d \vee \neg f_j) \tag{5}$$

$$\bigvee_{j \in K} \bigvee_{d \in D_{w,\overrightarrow{q_1 q_j}}} aux_{w,j,d} \tag{6}$$

3. For each $w \in S^-$ and each q_j, either there is no path state q_1 to q_j, or q_j is not final:

$$\neg \left[\bigvee_{j \in K} \bigvee_{d \in D_{w,\overrightarrow{q_1 q_j}}} (d \wedge f_j) \right] \tag{7}$$

Thus, the direct constraint model DM_k for building a NFA of size k is:

$$DM_k = \bigwedge_{w \in S^+} \left((4) \wedge (5) \wedge (6) \right) \wedge \bigwedge_{w \in S^-} (7)$$

and is possibly completed by (1) or (2) if $\lambda \in S^+$ or $\lambda \in S^-$.

Size of the Models *(see [9] for details).* Consider ω_+ and ω_-, the longest word of S^+ and S^- respectively. Table 1 presents the number of clauses (Column 1) and their arities (Column 2), which are an upper bound of a given constraint group (last column) for the model SM_k. Table 2 presents the upper bound of the number of Boolean variables that are required and why the are required. We can see on Tables 1 and 2 that the space complexity of the DM_k is huge ($\mathcal{O}(|S^+|.k.^{|\omega_+|})$ variables, and $\mathcal{O}(|S^+|.(|\omega_+|+1).k^{|\omega_+|})$ clauses) and with large clauses (up to arity of $|\omega_+|+2$), and that only small instances for a small number of states will be tractable. It is thus obvious that it is important to improve the model DM_k.

2.2 Prefix Model [2]

Let $Pref(w)$ be the set of all the non-empty prefixes of the word w and, by extension, $Pref(W) = \cup_{w \in W} Pref(w)$ the set of prefixes of the words of the set W. For each $w \in Pref(S)$, we add a Boolean variable $p_{w,\overrightarrow{q_1 q_i}}$ which determines whether there is or not a c_path for w from state q_1 to q_i. Note that these variables can be seen as labels of the Prefix Tree Acceptor (PTA) for S [3]. The problem can be modeled with the following constraints:

1. For all prefix $w = a$ with $w \in Pref(S)$, and $a \in \Sigma$, there is a c_path of size 1 for w:

$$\bigvee_{i \in K} \delta_{a,\overrightarrow{q_1 q_i}} \leftrightarrow p_{a,\overrightarrow{q_1 q_i}} \tag{8}$$

With the Tseitin transformations, we can derive a CNF formula. It is also possible to directly encode $\delta_{a,\overrightarrow{q_1 q_i}}$ and $p_{a,\overrightarrow{q_1 q_i}}$ as the same variable. Thus, no clause is required.

2. For all words $w \in S^+ - \{\lambda\}$:

$$\bigvee_{i \in K} p_{w,\overrightarrow{q_1 q_i}} \wedge f_i \tag{9}$$

With the Tseitin transformations [14], we create one auxiliary variable for each combination of $p_{w,\overrightarrow{q_1 q_i}}$ and the status (final or not) of the state q_i: $aux_{w,i} \leftrightarrow p_{w,\overrightarrow{q_1 q_i}} \wedge f_i$. Hence, for each w, we obtain a formula in CNF:

$$\bigwedge_{i \in K} ((\neg aux_{w,i} \vee p_{w,\overrightarrow{q_1 q_i}}) \wedge (\neg aux_{w,i} \vee f_i)) \tag{10}$$

$$\bigwedge_{i \in K} (aux_{w,i} \vee \neg p_{w,\overrightarrow{q_1 q_i}} \vee \neg f_i) \tag{11}$$

$$\bigvee_{i \in K} aux_{w,i} \tag{12}$$

3. For all words $w \in S^- - \{\lambda\}$, we obtain the following CNF constraint:

$$\bigwedge_{i \in K} (\neg p_{w,\overrightarrow{q_1 q_i}} \vee \neg f_i) \tag{13}$$

4. For all prefix $w = va$, $w \in Pref(S)$, $v \in Pref(S)$ and $a \in \Sigma$:

$$\bigwedge_{i \in K} (p_{w,\overrightarrow{q_1 q_i}} \leftrightarrow (\bigvee_{j \in K} p_{v,\overrightarrow{q_1 q_j}} \wedge \delta_{a,\overrightarrow{q_j q_i}})) \tag{14}$$

Applying the Tseitin transformations, we create one auxiliary variable for each combination of existence of a c_path from q_1 to q_i ($p_{v,\overrightarrow{q_1 q_i}}$) and the transition $\delta_{a,\overrightarrow{q_j q_i}}$: $aux_{v,a,j,i} \leftrightarrow p_{v,\overrightarrow{q_1 q_j}} \wedge \delta_{a,\overrightarrow{q_j q_i}}$. Then, (14) becomes:

$$\bigwedge_{i \in K} (p_{w,\overrightarrow{q_1 q_i}} \leftrightarrow (\bigvee_{j \in K} aux_{v,a,j,i}))$$

For each $w \in Pref(S)$, we obtain constraints in CNF:

$$\bigwedge_{(i,j)\in K^2} (\neg aux_{v,a,j,i} \vee p_{w,\overrightarrow{q_1q_i}}) \tag{15}$$

$$\bigwedge_{(i,j)\in K^2} (\neg aux_{v,a,j,i} \vee \delta_{a,\overrightarrow{q_jq_i}}) \tag{16}$$

$$\bigwedge_{(i,j)\in K^2} (aux_{v,a,j,i} \vee \neg p_{w,\overrightarrow{q_1q_i}} \vee \neg \delta_{a,\overrightarrow{q_jq_i}}) \tag{17}$$

$$\bigwedge_{i\in K} (\neg p_{w,\overrightarrow{q_1q_i}} \vee (\bigvee_{j\in K} aux_{v,a,j,i})) \tag{18}$$

$$\bigwedge_{(i,j)\in K^2} (p_{w,\overrightarrow{q_1q_i}} \vee \neg aux_{v,a,j,i})) \tag{19}$$

Thus, the constraint prefix model PM_k for building a NFA of size k is:

$$PM_k = \bigwedge_{w\in S^+} \left((10)\wedge\ldots\wedge(12)\right) \wedge \bigwedge_{w\in S^-} (13) \wedge \bigwedge_{w\in Pref(S)} (15)\wedge\ldots\wedge(19)$$

and is possibly completed by (1) or (2) if $\lambda \in S^+$ or $\lambda \in S^-$.

Size of the Models. Consider ω_+, the longest word of S^+, ω_-, the longest word of S^-, $\sigma = \Sigma_{w\in S}|w|$, and π, the number of prefix obtained by $Pref(S)$ with a size larger than 1 ($\pi = |\{x|x \in Pref(S), |x| > 1\}|$), then:

$$max(|\omega_+|,|\omega_-|) \leq \pi \leq \sigma \leq |S^+|.|\omega_+| + |S^-|.|\omega_-|$$

The space complexity of the PM_k model is thus in $\mathcal{O}(\sigma.k^2)$ variables, and in $\mathcal{O}(\sigma.k^2)$ binary and ternary clauses, and $\mathcal{O}(\sigma.k)$ $(k+1)$-ary clauses (Tables 3 and 4).

2.3   Suffix Model

We now propose a suffix model (SM_k), based on $Suf(S)$, the set of all the non-empty suffixes of all the words in S. The main difference is that the construction starts from every state and terminates in state q_1. For each $w \in Suf(S)$, we add a Boolean variable $p_{w,\overrightarrow{q_iq_j}}$ which determines whether there is or not a c-path for w from state q_i to q_j. To model the problem, Constraints (10), (11), (12), and (13) remain unchanged and creation of the corresponding auxiliary variables $aux_{w,i}$ as well.

Table 3. Clauses for PM_k

Number of cl.	Arity	Constraints
$2.k.|S^+|$	2	(10)
$k.|S^+|$	3	(11)
k	$k+1$	(12)
$k.|S^-|$	2	(13)
$\pi.k^2$	2	(15)
$\pi.k^2$	2	(16)
$\pi.k^2$	3	(17)
$\pi.k$	$k+1$	(18)
$\pi.k^2$	2	(19)

Table 4. Variables for PM_k

Number of var	Reason
k	Final states F
$n.k^2$	Transitions δ
$|S^+|.k$	Constraints (9)
$\pi.k^2$	Constraints (14)

For each suffix $w = a$ with $w \in Suf(S)$, and $a \in \Sigma$, there is a c_path of size 1 for w:

$$\bigvee_{(i,j)\in K^2} \delta_{a,\overrightarrow{q_i q_j}} \leftrightarrow p_{a,\overrightarrow{q_i q_j}} \tag{20}$$

We can directly encode $\delta_{a,\overrightarrow{q_i q_j}}$ and $p_{a,\overrightarrow{q_i q_j}}$ as the same variable. Thus, no clause is required.

For all suffix $w = av$, $w \in Suf(S)$, $v \in Suf(S)$ and $a \in \Sigma$:

$$\bigwedge_{(i,j)\in K^2} (p_{w,\overrightarrow{q_i q_j}} \leftrightarrow (\bigvee_{k\in K} \delta_{a,\overrightarrow{q_i q_k}} \wedge p_{v,\overrightarrow{q_k q_j}})) \tag{21}$$

We create one auxiliary variable for each combination of existence of a c_path from q_k to q_j ($p_{v,\overrightarrow{q_k q_j}}$) and the transition $\delta_{a,\overrightarrow{q_i q_k}}$: $aux_{v,a,i,k,j} \leftrightarrow \delta_{a,\overrightarrow{q_i q_k}} \wedge p_{v,\overrightarrow{q_k q_j}}$
For each $w = av$, we obtain the following constraints (CNF formulas):

$$\bigwedge_{(i,j,k)\in K^3} (\neg aux_{v,a,i,k,j} \vee p_{w,\overrightarrow{q_k q_j}}) \tag{22}$$

$$\bigwedge_{(i,j,k)\in K^3} (\neg aux_{v,a,i,k,j} \vee \delta_{a,\overrightarrow{q_i q_k}}) \tag{23}$$

$$\bigwedge_{(i,j,k)\in K^3} (aux_{v,a,i,k,j} \vee \neg p_{w,\overrightarrow{q_k q_j}} \vee \neg \delta_{a,\overrightarrow{q_i q_k}}) \tag{24}$$

$$\bigwedge_{(i,j)\in K^2} (\neg p_{w,\overrightarrow{q_i q_j}} \vee (\bigvee_{k\in K} aux_{v,a,i,k,j})) \tag{25}$$

Table 5. Clauses for SM_k

Number of cl.	Arity	Constraints
$2.k.|S^+|$	2	(10)
$k.|S^+|$	3	(11)
k	$k+1$	(12)
$k.|S^-|$	2	(13)
$\pi.k^3$	2	(22)
$\pi.k^3$	2	(23)
$\pi.k^3$	3	(24)
$\pi.k^2$	$k+1$	(25)
$\pi.k^3$	2	(26)

Table 6. Variables for SM_k

Number of var	Reason
k	Final states F
$n.k^2$	Transitions δ
$|S^+|.k$	Constraints (9)
$\pi.k^3$	Constraints (21)

$$\bigwedge_{(i,j,k)\in K^3} (p_{w,\overrightarrow{q_i q_j}} \vee \neg aux_{v,a,i,k,j})) \tag{26}$$

Note that some clauses are not worth being generated. Indeed, it is useless to generate paths starting in states different from the initial state q_1, except when the w is in S, and w is also the suffix of another word from S. Removing these constraints does not change the complexity of the model. This can easily be done at generation time, or we can leave it to the solver, which will detect it and remove the useless constraints.

Thus, the constraint prefix model PM_k for building a NFA of size k is:

$$SM_k = \bigwedge_{w\in S^+} \left((10)\wedge\ldots\wedge(12)\right) \wedge \bigwedge_{w\in S^-} (13) \wedge \bigwedge_{w\in Pref(S)\backslash S} (22)\wedge\ldots\wedge(26)$$

and is possibly completed by (1) or (2) if $\lambda \in S^+$ or $\lambda \in S^-$.

Size of the Models. Consider ω_+, ω_-, σ, and π as defined in the prefix model. Table 5 presents the number of clauses (first column) and their arities (Column 2) which are an upper bound of a given constraint group (last column) for the model SM_k. Table 6 presents the upper bound of the number of Boolean variables that are required, and the reason of their requirements. To simplify, the space complexity of SM_k is thus in $\mathcal{O}(\sigma.k^3)$ variables, and in $\mathcal{O}(\sigma.k^3)$ binary and ternary clauses, and $\mathcal{O}(\sigma.k^2)$ $(k+1)$-ary clauses.

3 Hybrid Models

We now propose a family of models based on both the notion of prefix and the notion of suffix. The idea is, in fact, to take advantage of the construction of a prefix p and a suffix s of a word w such that $w = p.s$ to pool both prefixes and suffixes. The goal is to reduce the size of generated SAT instances. The process is the following:

1. For each word w_i of S, we split w_i into p_i and s_i such that $w = p_i.s_i$. We thus obtain two sets, $S_p = \{p_i \mid \exists i, w_i \in S \text{ and } w_i = p_i.s_i\}$ and $S_s = \{s_i \mid \exists i, w_i \in S \text{ and } w_i = p_i.s_i\}$.
2. We then consider S_p as a sample, i.e., a set of words. For each w of S_p, we generate Constraints (15) to (19).
3. We consider S_s in turn to generate Constraints (22) to (26) for each $w \in S_s$.
4. Then, for each $w_i = p_i.s_i$, clauses corresponding to p_i must be linked to clauses of s_i.
   - if $w_i = p_i.s_i \in S^-$, the constraints are similar to the ones of (13) including the connection of p_i and s_i:

$$\bigwedge_{(j,k)\in K^2} (\neg p_{p_i,\overrightarrow{q_1 q_j}} \vee \neg p_{s_i,\overrightarrow{q_j q_k}} \vee \neg f_i) \tag{27}$$

   - if $w_i = p_i.s_i \in S^+$, the constraints are similar to (9):

$$\bigvee_{(j,k)\in K^2} p_{p_i,\overrightarrow{q_1 q_j}} \wedge p_{s_i,\overrightarrow{q_j q_k}} \wedge f_k \tag{28}$$

We transform (28) using auxiliary variables $aux_{w_i,j,k} \leftrightarrow p_{w,\overrightarrow{q_1 q_j}} \wedge p_{w,\overrightarrow{q_j q_k}} \wedge f_i$ to obtain the following CNF constraints:

$$\bigwedge_{(j,k)\in K^2} ((\neg aux_{w_i,j,k} \vee p_{w,\overrightarrow{q_1 q_j}}) \wedge (\neg aux_{w_i,j,k} \vee p_{w,\overrightarrow{q_j q_k}}) \wedge (\neg aux_{w_i,j,k} \vee f_k)) \tag{29}$$

$$\bigwedge_{(j,k)\in K^2} (aux_{w_i,j,k} \vee \neg p_{w,\overrightarrow{q_1 q_j}} \vee p_{w,\overrightarrow{q_j q_k}} \vee \neg f_k) \tag{30}$$

$$\bigvee_{(j,k)\in K^2} aux_{w_i,j,k} \tag{31}$$

Thus, the hybrid model HM_k for building a NFA of size k is:

$$HM_k = \bigwedge_{w \in S^+} \left((29) \wedge \ldots \wedge (31) \right) \wedge \bigwedge_{w \in S^-} (27) \wedge \bigwedge_{p_i \in Pref(S_p)} (22) \wedge \ldots \wedge (26) \bigwedge_{s_i \in Suf(S_s)} (15) \wedge \ldots \wedge (19)$$

and it is possibly completed by (1) or (2) if $\lambda \in S^+$ or $\lambda \in S^-$.

We do not detail it here, but in the worst case, the complexity of the model is the same as SM_k. It is obvious that the split of each word into a prefix and a suffix will determine the size of the instance. The next sub-sections are dedicated to the computation of this separation $w_i = p_i.s_i$ to minimize the size of the generated hybrid instances with the HM_k model.

3.1 Search Space and Evaluation Function for Metaheuristics

The search space $\mathcal{X}$ of this problem corresponds to all the hybrid models: for each word w of S, we have to determine a n such that $w = p.s$ with $|p| = n$ and $|s| = |w| - n$. The size of the search space is thus: $|\mathcal{X}| = \Pi_{w \in S}|w| + 1$.

Even though we are aware that smaller instances are not necessarily easier to solve, we choose to define the first evaluation function as the number of generated SAT variables. However, this number cannot be computed a priori: first, the instance has to be generated, before counting the variables. This function being too costly, we propose an alternative evaluation function for approximating the number of variables. This fitness function is based on the number of prefixes in $Pref(S_p)$ and suffixes in $Suf(S_s)$. Since the complexity of SM_k is in $\mathcal{O}(k^3)$ whereas the complexity of PM_k is in $\mathcal{O}(k^2)$, suffixes are penalized by a coefficient corresponding to the number of states.

$$fitness(S_p, S_s) = |Pref(S_p)| + k.|Suf(S_s)|$$

Empirically, we observe that the results of this *fitness* function are proportional to the actual number of generated SAT variables. This approximation of the number of variables will thus be the fitness function in our ILS and GA algorithms.

3.2 Iterated Local Search Hybrid Model HM_ILS_k

We propose an Iterated Local Search (ILS) [15] for optimizing our hybrid model. Classically, a best improvement or a first improvement neighborhood is used in ILS to select the next move. In our case, a first improvement provides very poor results. Moreover, it is clearly impossible to evaluate all the neighbors at each step due to the computing cost. We thus decide to randomly choose a word in S with a roulette wheel selection based on the word weights. Each word w has a weight corresponding for 75% to a characteristic of S, and 25% to the length of the word:

$$\text{weight}_w = 75\%/|S| + 25\% * |w|/(\sum_{w_i \in S} |w_i|)$$

The search starts generating a random couple of prefixes and suffixes sets (S_p, S_s), i.e., for each word w of S an integer is selected for splitting w into a prefix p and a suffix s such that $w = p.s$. Hence, at each iteration, the best couple (p, s) is found for the selected word w. This process is iterated until a maximum number of iterations is reached.

In our ILS, it is not necessary to introduce noise with random walks or restarts because our process of selection of word naturally ensures diversification.

Algorithm 1: Iterated Local Search

Input: set of words S, maximum number of iterations max_iter,
maximum of consecutive iterations allowed without improvement
$max\_iter\_without\_improv$

Output: set of prefixes S_p^*, set of suffixes S_s^*

1: Couple of prefixes and suffixes sets (S_p, S_s) is randomly generated
2: $(S_p^*, S_s^*) = (S_p, S_s)$
3: **repeat**
4:     Choose a word w in S with a roulette wheel selection
5:     (S_p, S_s) is updated by the best couple of the sub-search space corresponding
       only to a modification of the prefix and the suffix of word w
6:     **if** $fitness(S_p, S_s) < fitness(S_p^*, S_s^*)$ **then**
7:         $(S_p^*, S_s^*) = (S_p, S_s)$
8:     **end if**
9: **until** maximum number of iterations max_iter is reached or (S_p^*, S_s^*) is not
   improved since $max\_iter\_without\_improv$ iterations
10: **return** (S_p^*, S_s^*)

3.3 Genetic Algorithm Hybrid Model HM_GA_k

We propose a classical genetic algorithm (GA) based on the search space and
fitness function presented in Sect. 3.1. A population of individuals, represented
by a couple of prefixes and suffixes sets, is improved generation after generation.
Each generation keeps a portion of individuals as parents and creates children
by crossing the selected parents. Crossover operator used in our GA is the well-
known uniform crossover. For each word, children inherit the prefix and the
suffix of one of their parents randomly chosen. Since the population size is the
same during all the search, we have a steady-state GA. A mutation process is
applied over all individuals with a probability p_{mut}. For each word w, each prefix
and suffix are randomly mutated by generating an integer n between 0 and $|w|$
splitting w into a new prefix of size n and a new suffix $|w| - n$. The search stops
when the maximum number of generations is reached or when no improvement
is observed in the population during $max\_gen\_without\_improv$ generations.

4 Experimental Results

To test our new models, we work on the training set of the StaMinA Competition
(see http://stamina.chefbe.net). We use 11 of the instances selected in [2][1] with
a sparsity $s \in \{12.5\%, 25\%, 50\%, 100\%\}$ and an alphabet size $|\Sigma| \in \{2, 5, 10\}$.
We try to generate SAT instances for NFA sizes (k) near to the threshold of the
existence or not of an NFA.

---

[1] We kept the "official" name used in [2].

Algorithm 2: Genetic Algorithm

Input: set of words S, population size $s_{\mathcal{P}}$, mutation probability p_{mut},
maximum number of generations max_gen,
portion of population conserve in the next generation $p_{parents}$,
maximum of consecutive generations allowed without improvement
$max\_gen\_without\_improv$

Output: set of prefixes S_p^*, set of suffixes S_s^*

1: Population $\mathcal{P}$ of couples of prefixes and suffixes sets (S_p, S_s) is randomly
   generated
2: $(S_p^*, S_s^*) = \text{Argmin}_{fitness}(\mathcal{P})$
3: **repeat**
4:    Select as parents set Par a portion $p_{parents}$ of $\mathcal{P}$
5:    Generate $(1 - p_{parents}).s_{\mathcal{P}}$ children by uniform crossover over parents in a
      set $Children$
6:    $\mathcal{P} = Par \cup Children$
7:    Mutate for each individual of $\mathcal{P}$ the prefix/suffix for each words of S with a
      probability p_{mut}
8:    Update the population
9:    Update (S_p^*, S_s^*) if necessary
10: **until** maximum number of generations max_gen is reached or (S_p^*, S_s^*) is not
    improved since $max\_gen\_without\_improv$ generations
11: **return** S_p^* and S_s^*

4.1 Experimental Protocol

All our algorithms are implemented in Python using specific libraries such as
Pysat. The experiments were carried out on a computing cluster with Intel-E5-
2695 CPUs, and a limit of 10 GB of memory was fixed. Running times were
limited to 10 min, including generation of the model and solving time. We used
the Glucose [13] SAT solver with the default options. For stochastic methods
(ILS and GA), 30 runs are realized to exploit the results statistically.

Parameters used for our hybrid models are:

ILS		AG	
max_iter	10 000	$s_{\mathcal{P}}$	100
$max\_iter\_without\_improv$	100	max_gen	3000
		$max\_gen\_without\_improv$	100
		p_{mut}	0.05
		$p_{parents}$	0.03

4.2   Results

Our experiments are reported in Table 7. The first column (*Instance*) corresponds to the official name of the instance, and the second one (k) to the number of states of the expected NFA. Then, we have in sequence the model name (*Model*), the number of SAT variables (*Var.*), the number of clauses (*Cl.*), and the instance generation time (t_M). The right part of the table corresponds to the solving part with the satisfiability of the generated instance (*SAT*), the decisions number (*Dec.*), and the solving time (t_S) with Glucose. Finally, the last column (t_T) corresponds to the total time (modeling time + solving time). Results for hybrid models based on ILS (HM_ILS_k) and GA (HM_GA_k) correspond to average values over 30 runs. We have decided to only provide the average since the standard deviation values are very small.

The last lines of the table correspond to the cumulative values for each column and each model. When an instance is not solved (time-out), the maximum value needed for solving the other model instances is considered. For the instance generation time (t_M), a credit of 600 s is applied when generation did not succeed before the time-out.

We can clearly confirm that the direct model is not usable in practice, and that instances cannot be generated in less than 600 s. The prefix model allows the fastest generation when it terminates before the time out (on these benchmarks, it did not succeed once and was thus penalize for cumulative values). It also provides instances that are solved quite fast. As expected, the instances optimized with GA are the smallest ones. However, the generation is too costly: the gain in solving time is not sufficient to compensate the long generation time. In total, in terms of solving+generation time, GA based model is close to prefix model. As planned with its space complexity (in $\mathcal{O}(k^3)$), suffix based instances are huge and long to solve. However, we were surprised for 2 benchmarks (ww-10-40 and ww-10-50) for which the generated instances are relatively big (5 times the size of the GA optimized instances), but their solving is the fastest. We still cannot explain what made these instances easy to solve, and we are still investigating their structure. The better balance is given with the ILS model: instances are relatively small, the generation time is fast, and the solving time as well. This is thus the best option of this work.

It is very difficult to compare our results with the results of [2]. First of all, in [2], they try to minimize k, the number of states. Moreover, they use parallel algorithms. Finally, they do not detail the results for each instance and each k, except for st-2-30 and st-5-50. For the first one, with $k = 9$ we are much faster. But for the second one, with $k = 5$ we are slower.

Table 7. Comparison on 11 instances between the models DM_k, PM_k, SM_k, HM_ILS_k, and HM_GA_k.

Instance	k	Model	Var.	Cl.	t_M	SAT	Dec.	t_S	t_T
st-2-10	4	DM_k	190 564	1 817 771	13.46	True	973 213	88.62	102.09
		PM_k	1 276	4 250	1.27	True	3 471	0.10	**1.37**
		SM_k	5 196	17 578	1.30	True	4 332	0.21	1.51
		HM_ILS_k	1 188	4 179	4.14	True	2 503	**0.05**	4.18
		HM_GA_k	**1 107**	3 884	14.45	True	2 368	**0.05**	14.50
st-2-20	6	DM_k	–	–	–	–	–	–	–
		PM_k	4 860	17 150	1,34	False	1 625 706	241,24	242,59
		SM_k	–	–	–	–	–	–	–
		HM_ILS_k	5 688	21 073	5,39	False	662 354	98,35	**103,74**
		HM_GA_k	**4 735**	17 611	34,65	False	708 356	**94,95**	129,61
st-2-30	9	DM_k	–	–	–	–	–	–	–
		PM_k	–	–	–	–	–	–	–
		SM_k	–	–	–	–	–	–	–
		HM_ILS_k	20 637	78 852	7.55	True	1 998 574	**228.53**	**236.07**
		HM_GA_k	**16 335**	62 832	66.94	True	4 079 686	527.44	594.38
st-5-20	4	DM_k	–	–	–	–	–	–	–
		PM_k	4 024	13 464	1.49	True	2 641	**0.08**	**1.57**
		SM_k	14 964	50 660	1.68	True	23 540	3.23	4.91
		HM_ILS_k	3 608	12 514	7.83	True	14 584	0.72	8.56
		HM_GA_k	**3 522**	12 180	47.84	True	18 344	0.94	48.78
st-5-30	4	DM_k	–	–	–	–	–	–	–
		PM_k	5 364	18 054	1.43	True	177 711	21.57	**23.00**
		SM_k	21 084	71 502	1.87	True	362 318	128.02	129.89
		HM_ILS_k	4 837	16 955	9.90	True	156 631	**19.90**	29.81
		HM_GA_k	**4 705**	16 478	119.42	True	171 062	21.67	141.09
st-5-40	4	DM_k	–	–	–	–	–	–	–
		PM_k	6 284	21 216	1.52	False	7 110	0.55	**2.07**
		SM_k	23 604	80 104	1.55	False	15 708	1.74	3.29
		HM_ILS_k	5 745	20 290	10.29	False	6 206	**0.34**	10.62
		HM_GA_k	**5 548**	19 517	150.50	False	6 204	0.35	150.85
st-5-50	5	DM_k	–	–	–	–	–	–	–
		PM_k	11 150	38 745	1.59	False	1 943 735	562.80	564.39
		SM_k	–	–	–	–	–	–	–
		HM_ILS_k	11 085	40 258	10.80	False	911 280	**238.10**	**248.90**
		HM_GA_k	**10 040**	36 350	279.87	False	1 093 093	287.46	567.33
st-5-60	5	DM_k	–	–	–	–	–	–	–
		PM_k	14 200	49 455	1.52	False	1 245 538	383.37	384.89
		SM_k	–	–	–	–	–	–	–
		HM_ILS_k	13 920	50 568	13.47	False	800 920	**231.82**	**245.29**
		HM_GA_k	**13 180**	47 755	313.30	False	950 601	270.97	584.26
ww-10-40	4	DM_k	15 012	112 039	2.07	True	69 219	1.52	3.59
		PM_k	3 624	11 900	1.38	True	977	0.03	1.41
		SM_k	13 844	46 648	1.25	True	4 173	**0.02**	**1.28**
		HM_ILS_k	2 896	10 342	5.94	True	3 897	0.06	6.00
		HM_GA_k	**2 761**	9 839	75.57	True	2 842	0.04	75.60
ww-10-50	4	DM_k	80 548	694 641	5.61	True	483 153	103.52	109.14
		PM_k	5 364	17 850	1.28	True	167 390	20.29	21.57
		SM_k	20 844	70 482	1.49	True	74 482	11.58	**13.07**
		HM_ILS_k	4 633	16 514	7.71	True	73 534	5.46	13.17
		HM_GA_k	**4 517**	15 940	123.21	True	52 894	**3.38**	126.59
Cumulative values		DM_k	397 451	3 014 842	4 221,15	–	10 821 816	2 041,50	6 262,65
		PM_k	76 783	270 936	612,82	–	9 253 965	1 757,47	2 370,29
		SM_k	151 211	525 099	2 409,15	–	9 379 218	1 879,65	4 268,80
		HM_ILS_k	74 237	271 543	**83,03**	–	4 630 483	**823,33**	**906,36**
		HM_GA_k	**66 450**	242 386	1 225,75	–	7 085 451	1 207,23	2 432,98

5   Conclusion

In this paper, we have proposed to use some metaheuristics algorithms, namely ILS and GA, to improve the size of SAT models for the NFA inferring problem. Our hybrid model, optimized with GA gives, on average, the smallest SAT instances. Solving these instances is also faster than with the direct or prefix models. However, generation of the optimized instances with GA is really too long and is not balanced out with the gain in solving time; it is at the level of the prefix model w.r.t. total CPU time. The ILS model generates optimized instances a bit larger than with GA and a bit smaller than with prefixes. Moreover, the solving time is the best of our experiments, and the generation time added to the solving time makes of the HM_ILS_k our better model.

In the future, we plan to speed up GA to make it more competitive. We also plan to consider more complex fitness functions, not only based on the number of SAT variables but also on the length of clauses. We also plan a model portfolio approach for larger samples.

References

1. Wieczorek, W.: Grammatical Inference - Algorithms, Routines and Applications. Studies in Computational Intelligence, vol. 673. Springer, Cham (2017). https://doi.org/10.1007/978-3-319-46801-3
2. Jastrzab, T., Czech, Z.J., Wieczorek, W.: Parallel algorithms for minimal nondeterministic finite automata inference. Fundam. Informaticae **178**, 203–227 (2021)
3. de la Higuera, C.: Grammatical Inference: Learning Automata and Grammars. Cambridge University Press, Cambridge (2010)
4. Denis, F., Lemay, A., Terlutte, A.: Learning regular languages using RFSAs. Theor. Comput. Sci. **313**, 267–294 (2004)
5. de Parga, M.V., García, P., Ruiz, J.: A family of algorithms for non determinestic regular languages inference. In: Ibarra, O.H., Yen, H.-C. (eds.) CIAA 2006. LNCS, vol. 4094, pp. 265–274. Springer, Heidelberg (2006). https://doi.org/10.1007/11812128_25
6. Tomita, M.: Dynamic construction of finite-state automata from examples using hill-climbing. In: Proceedings of the Annual Conference of the Cognitive Science Society, pp. 105–108 (1982)
7. Dupont, P.: Regular grammatical inference from positive and negative samples by genetic search: the GIG method. In: Carrasco, R.C., Oncina, J. (eds.) ICGI 1994. LNCS, vol. 862, pp. 236–245. Springer, Heidelberg (1994). https://doi.org/10.1007/3-540-58473-0_152
8. Rossi, F., van Beek, P., Walsh, T. (eds.): Handbook of Constraint Programming, 1st edn. Elsevier, Amsterdam (2006)
9. Lardeux, F., Monfroy, E.: Improved SAT models for NFA learning. In: Dorronsoro, B., Amodeo, L., Pavone, M., Ruiz, P. (eds.) OLA 2021. CCIS, vol. 1443, pp. 267–279. Springer, Cham (2021). https://doi.org/10.1007/978-3-030-85672-4_20
10. Garey, M.R., Johnson, D.S.: Computers and Intractability, A Guide to the Theory of NP-Completeness. W.H. Freeman & Company, San Francisco (1979)

11. Jastrzab, T.: Two parallelization schemes for the induction of nondeterministic finite automata on PCs. In: Wyrzykowski, R., Dongarra, J., Deelman, E., Karczewski, K. (eds.) PPAM 2017. LNCS, vol. 10777, pp. 279–289. Springer, Cham (2018). https://doi.org/10.1007/978-3-319-78024-5_25
12. Heule, M., Verwer, S.: Software model synthesis using satisfiability solvers. Empir. Softw. Eng. **18**, 825–856 (2013)
13. Audemard, G., Simon, L.: Predicting learnt clauses quality in modern SAT solvers. In: Proceedings of IJCAI 2009, pp. 399–404 (2009)
14. Tseitin, G.S.: On the complexity of derivation in propositional calculus. In: Siekmann, J.H., Wrightson, G. (eds.) Automation of Reasoning, pp. 466–483. Springer, Heidelberg (1983). https://doi.org/10.1007/978-3-642-81955-1_28
15. Stützle, T., Ruiz, R.: Iterated local search. In: Martí, R., Pardalos, P.M., Resende, M.G.C. (eds.) Handbook of Heuristics, pp. 579–605. Springer, Cham (2018). https://doi.org/10.1007/978-3-319-07124-4_8

Multi Phase Methodology for Solving the Multi Depot Vehicle Routing Problem with Limited Supply Capacity at the Depots

Javier de Prado, Sandro Moscatelli, Pedro Piñeyro⬤, Libertad Tansini$^{(\boxtimes)}$⬤, and Omar Viera

Department of Operations Research, Computer Science Institute (InCo), Faculty of Engineering, Universidad de la República, Montevideo, Uruguay
{moscatel,ppineyro,libertad,viera}@fing.edu.uy

Abstract. This paper focuses on a capacitated multi depot vehicle routing problem, where each depot has a finite supply capacity to meet the customers demand. To solve this problem we propose a multi phase methodology, that extends the "cluster first, route second" approach. It is based on iterative routings to find and reassign misplaced customers with respect to the depots and with the objective of improving the final routing. Several assignment and routing algorithms are considered to evaluate the proposed methodology under different settings. A mathematical model of the problem is given to perform a comparative study of the methodology against an exact solution method. The results obtained from the numerical experiments carried out allow us to conclude that the methodology can be successfully applied to the capacitated multi depot vehicle routing problem.

Keywords: Multi depot vehicle routing problem · Heuristics · Supply capacity · Clustering · Assignment

1 Introduction and Related Works

We address the problem of distribution of goods from several depots to a set of geographically dispersed customers with known coordinates and demand, assuming finite supply capacity at each depot and an unlimited fleet of homogeneous and capacitated vehicles. We refer to this problem as the Capacitated Multi-Depot Vehicle Routing Problem (CMDVRP). The objective is to determine a set of routes starting and ending at each depot, minimizing the total distance traveled and subject to the supply capacity of each depot and the capacities of the vehicles. The CMDVRP can be found in recent real life applications such as emergency facilities location-routing and city logistics problems [20,22]. The CMDVRP is an NP-hard problem since it can be considered an extension of the

B. Dorronsoro et al. (Eds.): META 2021, CCIS 1541, pp. 198–211, 2022.
https://doi.org/10.1007/978-3-030-94216-8_15

Multi-Depot Vehicle Routing Problem (MDVRP), which in turn is an extension of the classical Vehicle Routing Problem (VRP) [8].

We present here a novel multi phase methodology to solve the CMDVRP inspired by the "cluster first, route second" approach. The initial phase consists of the assignment of costumers to depots and the final phase produces the routing of the VRPs related to all depots. Between these two phases, there is an intermediate phase for the reassignment of customers to depots with the aim to obtain a high quality solution in the cluster first part, improving in this way the final routing phase. The detection and reassignment of customers are based on a combination of misplaced-customer criterion and routing algorithm. A misplaced customer is reassigned to another depot, if this reassignment improves the cost of the general solution, which is the objective of the proposed methodology. The main idea behind the proposed methodology is that the complexity of the algorithms used in each phase can be chosen by the decision makers according to their needs and possibilities. The strength of the methodology is to provide good quality solutions in reasonable times, even in the case of using simple algorithms (easy to understand and code).

As far as we know, only few authors focus on the CMDVRP, and in particular, by means of the "cluster first, route second" approach. Giosa et al. [9] describe and compare several assignment algorithms for the clustering phase. Tansini et al. [17] compare the results obtained by a set of heuristic algorithms for the assignment of customers to depots with assignments obtained from solving the Transport Problem. Six heuristics for the clustering problem (assignment of customers to depots) are presented and analyzed in [10]. Also [18] consider this approach for the real-life problem of milk collection. Allahyari et al. [1] tackle the CMDVRP extension in which every customer is satisfied either by visiting the customer or by being located within an acceptable distance from at least one visited customer. Calvet et al. [4] consider the CMDVRP for the case of customers with stochastic demand and supply constraints on the depots due to the limited number of capacitated vehicles assigned to each of them. A collaborative routing problem with shared carriers and multiple depots (wholesalers) with limited storage is tackled in [21].

We note that many authors have considered the multi-depot vehicle routing problem with limited capacity on vehicles and/or route lengths, but not on the supply depots. For instance, Vidal et al. [19] propose a framework to solve the MDVRP, the Periodic VRP (PVRP), and the multi-depot periodic VRP with capacitated vehicles and constrained route duration. Contardo and Martinelli [6] suggest an exact algorithm for the MDVRP under capacity and route length constraints, exploiting the vehicle-flow and set-partitioning formulations. Recently, Pessoa et al. [15] propose a generic branch-cut-and-price solver for different vehicle routing variants and related problems.

The remainder of this paper is organized as follows. In Sect. 2 we introduce the mathematical formulation for the CMDVRP. The proposed methodology for solving the CMDVRP is further described in Sect. 3. In Sect. 4 we present the results of the comparison between the methodology against exact methods and we also analyze the effectiveness of the exploration phase of the methodology. Finally, in Sect. 5, we provide the conclusions and some directions for future research.

2   The Capacitated Multi-Depot Vehicle Routing Problem (CMDVRP)

The CMDVRP can be formally described as follows, extending that presented in [14] for the MDVRP. Let $G = (V, E)$ be a directed graph, where V denotes the set of nodes $\{1, ..., n\}$ and $E \subseteq V \times V$ the set of arcs. Let D be the set of depot nodes $\{1, ..., m\}$, with $1 \leq m < n$, and U the set of customer nodes $\{(m+1), ..., n\}$. For each node $i \in V$ there is a related quantity $q_i \geq 0$ that represents either the supply capacity for nodes $i \in D$ or the demand requirements in the case of nodes $i \in U$. For each arc $(i, j) \in E$ there is a routing cost $c_{i,j} \geq 0$. Let also consider the set of the possible routes $R = \{1, ..., (n - m)\}$. A route r can be defined as either the empty set or a finite sequence of at least three elements of V satisfying the following conditions: 1) in the extremes there is the same node i, with $i \in D$, 2) the internal nodes are customers nodes j with $j \in U$, and 3) for any pair of nodes $j, k \in U$, we have that $j \neq k$. We assume that for each route r there is a vehicle of capacity $p \geq 0$. Then, the objective is to determine the set of routes r in R in order to fulfill the demand of each customer without exceeding the vehicle and depot capacities, minimizing the total cost of routing. To formulate the CMDVRP as a Mixed Integer Linear Programming (MILP) we define the binary variables x_{ijkr} to be equal to 1 only if the arc (i, j) is in the route r of the depot k; 0 otherwise. Thus, the MILP proposed for the CMDVRP is as follows:

$$\min \sum_{i \in V} \sum_{j \in V} \sum_{k \in D} \sum_{r \in R} c_{ij} x_{ijkr} \tag{1}$$

subject to:

$$\sum_{j \in V} \sum_{k \in D} \sum_{r \in R} x_{ijkr} = 1, \qquad \forall i \in U \tag{2}$$

$$\sum_{j \in V \setminus \{i\}} x_{ijkr} = \sum_{j \in V \setminus \{i\}} x_{jikr}, \qquad \forall i \in V, \forall k \in D, \forall r \in R \tag{3}$$

$$\sum_{i \in U} \sum_{j \in V \setminus \{i\}} \sum_{k \in D} q_i x_{ijkr} \leq p, \qquad \forall r \in R \tag{4}$$

$$\sum_{i \in U} \sum_{j \in V \setminus \{i\}} \sum_{r \in R} q_i x_{ijkr} \leq q_k, \qquad \forall k \in D \tag{5}$$

$$\sum_{j \in U} x_{kjkr} \leq 1, \qquad \forall k \in D, \forall r \in R \tag{6}$$

$$\sum_{j \in U} x_{ijkr} = 0, \qquad \forall i, k \in D, i \neq k, \forall r \in R \tag{7}$$

$$y_i - y_j + (n - m) \sum_{k \in D} \sum_{r \in R} x_{ijkr} \leq n - m - 1, \qquad \forall i, j \in U, i \neq j \tag{8}$$

$$y_i \geq 0, \qquad \forall i \in U \tag{9}$$

$$x_{ijkr} \in \{0, 1\}, \qquad \forall i, j \in V, \forall k \in D, \forall r \in R \tag{10}$$

The objective function (1) is the minimization of the total cost of distance traveled. Constraints (2) state that each customer is included in a single route. Constraints (3) are for the route continuity. Constraints (4) and (5) represent the vehicle and depot capacity, respectively. Constraints (6) and (7) state that one route is assigned at most to a single depot. In (8) are the constraints of Miller-Tucker-Zemlin for subtours elimination [3]. Finally, constraints (9) and (10) are for the domain of values of the decision variables.

Although the main difference between CMDVRP and MDVRP are the constraints of (5), we note that, in general, a more restricted problem is more difficult to solve.

3   Multi-phase Methodology for the CMDVRP

It is worth to note that the assignment problem and the routing problem in the "cluster first, route second" approach are not independent from each other. A bad assignment solution will result in routes of higher total cost, even if an effective routing algorithm is used. Motivated by this, we consider an improvement to this approach, by means of a multi-phase methodology (MPM) for solving the CMDVRP. It begins from an initial assignment of costumers to depots and in the final phase produces the routing of the VRPs related to all depots. We introduce an intermediate phase in which misplaced costumers are detected and may be reassigned to another depot, if it improves the cost of the overall solution. Successive reassignment of misplaced costumers, based on the routing, will in most cases lead to an improvement of the solution. An outline of the proposed methodology for the CMDVRP is as follows:

1. **Assignment phase**: choose and apply an assignment algorithm of customers to depots taking into account demand and supply restrictions. The choice may depend on computational time and other restrictions.
2. **Exploration phase**: choose and apply a routing algorithm for all VRPs related to the depots. Again, the choice may depend on computational time and other restrictions. Then, repeat until no further improvement can be achieved:

(a) Detect misplaced customers based on the current assignment and eventually other restrictions.
(b) Reassign misplaced customers and run the selected routing algorithm. Accept the reassignment if it improves the cost of the overall solution.
3. **Final routing phase**: choose and apply the final routing algorithm for all VRPs related to the depots.

One of the advantages of the suggested MPM is that each phase offers the possibility of choosing different algorithms depending on the specific characteristics of the problem, the problem instances, hardware limitations, time restrictions, etc. They can be exchanged and combined in different manners. Thus, a specific selection of algorithms for each phase produces a particular MPM instantiation that can be considered a heuristic procedure to solve the CMDVRP. Next the MPM phases are explained in more detail and some algorithms that can be used in each one are mentioned.

3.1   Assignment Phase

Since each phase of the methodology offers a great variety of possibilities to instantiate and since there are several known methods that can be used to obtain an initial assignment of customers, in this work we narrow down the study to two assignment schemes.

We use the urgency assignment (Ur) which is a simple and fast assignment method that considers an urgency value μ_c for each customer c that determines the order in which customers are assigned to depots with limited supply capacity [9], as follows:

$$\mu_c = \left[\sum_{d \in D} dist(c, d) \right] - dist(c, d') \tag{11}$$

where $dist(c, d)$ is the distance of customer c to depot d, and $dist(c, d')$ is the distance to the closest depot d'. This measure accounts for the cost of assigning a customer to a depot other than its closest depot. Customers with more urgency (higher μ_c value) will be assigned first. Once a depot is complete it will no longer be considered for the further assignments and will hence not participate in the urgency calculations. Note that after each assignment the urgency of some customers must be recalculated.

Alternatively in this study the modified urgency assignment (MUr) is used as another assignment method and is defined as the combination of the urgency assignment [9] and the cluster assignment [10]. Customers are assigned to depots with the same criterion as in the urgency assignment until a fraction of them have been assigned (in this case 1/4) and then finalizes by assigning customers to the closest cluster made up of each depot and the already assigned customers, where it is feasible to assign the customer, i.e. will not exceed the total capacity of the depot.

There are other interesting assignment algorithms that could be explored such as the sweep approach [11] or using a grid or Voronoi diagrams [2]. We note that some of them do not consider the capacity of the depots and may require an adaptation or post-processing in order to give an acceptable initial assignment.

3.2   Exploration Phase

The exploration phase is the keystone of the proposed methodology. It is characterized by: 1) the definition of misplaced customers; 2) the processing order of the misplaced customers; 3) the routing algorithm used iteratively; 4) the criterion that determines if each misplaced customer should be reassigned or not; and 5) the reassignment strategy. In the following sections we describe the definitions and algorithms to be used in our study for this phase.

Definition of Misplaced Customers: Here, the definition of misplaced customers, the processing order of the misplaced customers, and the criterion that determines if each misplaced customer should be reassigned or not, all of them depend on the routing algorithm that is used iteratively to obtain the results of the VRPs related to all depots.

It is possible to infer that different definitions of misplaced customers lead to different ways of exploring neighboring solutions. The first approach was to define misplaced customers as those whose two closest customers are assigned to another depot. In general, we can define a misplaced customer as that for which its n closest customers are assigned to other depots (possibly different), for certain positive integer $n > 0$. Thus, a more flexible definition considers a customer to be misplaced if considering its n closest customers, m of them are assigned to other depots, where $m \leq n$. Observe that this definition focuses on the cost of the solution, therefore the reassignment strategy considers the supply capacity of the depots.

Other approaches would be to consider constraints such as capacity and time windows in the definition of misplaced customers.

Processing Order of the Misplaced Customers: In this work, misplaced customers are processed in descending order of the following misplaced criterion:

$$\varphi_c = dist(c, d) - \left[\sum_{i=1}^{N} dist(c, c_i) \right] \tag{12}$$

where customer c has been assigned to the depot d and $c_1, ..., c_N$ are its N closest customers not assigned to d, but assigned all to the same depot. The value of φ_c can be positive or negative, where a high positive value of φ_c means that the customer c is very far from the assigned depot compared to the distance to its closest neighbors.

Routing Algorithm Used Iteratively: Three different algorithms for the routing of the customers assigned to each depot were tested in this paper for the Exploration phase: Clarke and Wright algorithm [5], Sweep [11] and JSprit (available at https://jsprit.github.io/) which is a metaheuristic defined by the ruin-and-recreate principle [16]. It is a highly optimized method that consists of a large neighborhood search that combines elements of simulated annealing and threshold-accepting algorithms. Both Clarke and Wright and Sweep are classical and simple routing algorithms for the VRP, and also Clarke and Wright is a very popular constructive heuristic [12] and Sweep is the most elementary version of petal-type constructive heuristics [12]. It is worth noting that in each iteration, the routing algorithm only needs to bee applied for those depots with reassigned customers, since the others remain unchanged.

Reassignment Criterion: The reassignment criterion used in this paper is to reassign a customer if it produces a lower routing cost than the current assignment. It is important to note that once the reassignments are decided, it is only necessary to run the routing algorithm for the implicated depots. The routing for the rest of the depots remains unchanged.

Reassignment Strategy: Different approaches can be considered for the reassignment strategy. They should describe the conditions and the procedure to assign a misplaced customer to another depot, that will potentially improve the final routing. In general, the demand of customers and the capacities of depots should be considered. In this paper the reassignment strategy is determined by a two-stage procedure executed over an ordered list of misplaced customers. As part of the strategy, it has to be decided the number m of customers with the same target depot that may be considered to be reassigned simultaneously. This section explains the strategy suggested to reassign one misplaced customer ($m = 1$) at a time in the exploration phase.

In the first stage, the reassignment of a misplaced customer i to the depot d' of the closest customer i' is attempted, if d' has enough spare capacity to serve costumer i. If the reassignment produces a better overall routing result, it is accepted and the list of misplaced customers is recalculated. If there is no improvement, the next misplaced customer in order of misplaced criterion is considered to be reassigned. If depot d' does not have enough spare capacity to serve costumer i, then i is reassigned to the closest depot d'' (if it exists) that does have enough spare capacity to serve it. Again, if the reassignment produces a better overall routing result, the reassignment is accepted and the list of misplaced customers is recalculated.

The aim of the second stage is to try reassign the misplaced customers that remain in the list after the first stage. In this stage the same processing is done with the misplaced customers as in the previous stage except in the way of determining the alternative depot d'' and the reassignment moves. Let us assume that depot d' does not have enough spare capacity to serve the misplaced customer i under consideration, with d' as in the first stage. Then, a misplaced customer i'' assigned to d' is determined, that could potentially be reassigned to another depot d'', with d'' the depot of the closest customer to i'', allowing d' to serve the customer i. If misplaced customer i'' exists, a double reassignment is attempted by means of assigning i'' to depot d'' and i to depot d'. If the double reassignment of customers produces a better overall routing result, the reassignment is accepted and the list of misplaced customers is recalculated.

The two stages described above, are repeated until there are no further misplaced customers (the list is empty) or no misplaced customer reassignment results in a cost improvement.

In the case of at least two misplaced customers ($m \geq 2$) being reassigned together, the procedure is similar but the destination depot has to have enough spare capacity to serve the set of misplaced customers under consideration.

3.3   Final Routing Phase

Several routing algorithms can be used to produce the final routing once the Exploration phase has finished. In this work the same three algorithms used in the Exploration phase were tested for the Final routing phase.

4   Evaluation of the Proposed Methodology

In this section we provide the results obtained from different numerical experiments of several MPM instantiations. Given the reasonable computational time observed for the MPM methodology, we performed all experiments with all combinations of assignment and routing algorithms for the phases of the methodology (assignment, exploration and final routing phases). The MPM instantiation with the best result obtained is shown in the tables (in the case of equal cost the one with the fastest time is chosen).

The mathematical model for the CMDVRP presented in Sect. 2 was coded in AMPL and solved with CPLEX 12.6.3.0 on a PC Intel Core i7, 16 CPUs, 64 GB of RAM (DDR4) and CentOS 7. The MPM instantiations were coded in Java and executed in a PC with Intel Xeon CPU E3-1220 V2, 4 GB of RAM and Windows 7.

4.1   Comparative Study with Exact Method

Solving the CMDVRP to optimality is extremely costly due to the computational complexity of the problem. Nevertheless, an important aspect of a comprehensive analysis for any proposed heuristic approach is to compare its results against exact methods both regarding objective values and running times.

In   https://www.fing.edu.uy/owncloud/index.php/s/XnvURwxKzQUaH1P it is available the benchmark set of instances used to compare different MPM instantiations against CPLEX. Table 1 presents the results obtained.

The first column of Table 1 provides the identification of the instances, with 20 nodes in total, 2 or 3 depots, and a sequential number. The capacity of each depot is in the range $[66, 125]$, and the vehicle capacity in the range $[50, 70]$. The sum of the customers demand is of 180 units for each instance. We note that the distribution of the customers and depots is based on real map coordinates on certain islands of the Pacific ocean. Columns 2 to 4 report the name of the MPM instantiation, the costs and the total running times (in seconds) of all phases of the methodology for each one of the instances in the benchmark set. The name of each MPM instantiation is composed by four terms separated by a simple dash: the assignment algorithm (Ur or MUr), the criterion for misplaced customers and the routing algorithms used for the exploration and final phases, respectively. For example, a misplaced criterion 5c1n2m means that 5 closest customers are considered for determining if certain customer is misplaced, at least 1 of them is assigned to another depot, and 2 can be reassigned simultaneously. For all the experiments performed, we consider 1 to 5 closest customers for the misplaced criterion and between 1 or 2 customers to be reassigned simultaneously. The algorithms used in each phase and the misplaced criteria of the MPM instantiations listed in Table 1 were those for which we obtained the best results in the experiments. Columns 5 reports the cost obtained from CPLEX with a running time limited to 3600 s (no significant improvements were noticed with higher running times). Last column 6 in Table 1 provides the percentage gap between the cost of the MPM instantiation and CPLEX, calculated as $100 * (MPM_{Cost} - CPLEX_{Cost})/CPLEX_{Cost}$.

Table 1. Comparison of results for MPM instantiations against CPLEX.

Instance	MPM instantation	Time	Cost	CPLEX cost	% Gap cost
20n2d01	MUr-5c1n2m-JSprit-JSprit	58.79	3064.9	3085.83	−0.68
20n2d02	Ur-5c1n2m-JSprit-JSprit	24.757	5726.34	5726.34	0.00
20n2d03	Ur-5c1n2m-JSprit-JSprit	22.591	224.18	229.92	−2.50
20n2d04	Ur-1c1n1m-Sweep-C&W	0.001	158.03	158.03	0.00
20n2d05	Ur-5c1n2m-JSprit-JSprit	45.336	354.39	361.26	−1.90
20n2d06	Ur-5c1n2m-Sweep-JSprit	0.266	5808.51	5808.51	0.00
20n2d07	MUr-5c1n2m-JSprit-JSprit	11.337	5873.72	5873.71	0.00
20n2d08	Ur-5c1n2m-JSprit-JSprit	37.826	5062.75	5062.75	0.00
20n2d09	Ur-5c1n2m-Sweep-JSprit	0.298	910.97	924.15	−1.43
20n2d10	Ur-1c1n1m-Sweep-C&W	0.001	292.26	292.26	0.00
20n3d01	Ur-1c1n1m-C&W-JSprit	0.354	2556.36	2726.02	−6.22
20n3d02	Ur-5c1n2m-Sweep-JSprit	0.248	159.84	159.84	0.00
20n3d03	Ur-1c1n1m-C&W-JSprit	0.329	123.98	123.98	0.00
20n3d04	Ur-1c1n1m-C&W-JSprit	0.354	4773.08	4973.73	−4.03
20n3d05	Ur-1c1n1m-C&W-JSprit	0.347	4576.78	4576.78	0.00
Average		13.522			−1.12

From the results in Table 1 we note that MPM outperforms CPLEX in 6 of the 15 instances, and achieves the same objective value in the remaining ones.

Thus, we can conclude that MPM is competitive with CPLEX because the gap error is always negative or zero and the running times of all the considered MPM instantiations are significantly lower than CPLEX (less than 60 s versus 3600 s). We also want to note that the most effective MPM instantiation considering both, costs and running times, is Ur-1c1n1m-C&W-JSprit, since it shows the two lowest percentage cost gaps and less than a half of a second of running time. This MPM instantation makes use of different algorithm approaches for the exploration and final routing phases. This seems to indicate that it would be enough to use a simple and fast algorithm for the exploration phase, and a good and eventually time consuming routing algorithm for the final phase.

4.2   Comparative Study with and Without Exploration Phase

A central part of the proposed methodology, is the intermediate exploration phase for the detection of misplaced customers and the reassignment of them to depots using a routing algorithm. In this section we analyze the impact of including the exploration phase in the methodology by means of a comparative study over ten large instances with different geographical characteristics, available also at the same web repository provided in Sect. 4.1. Some of them are based on instances of the TSPLIB (http://comopt.ifi.uni-heidelberg.de/software/TSPLIB95/vrp/), and others were randomly generated, trying to create clusters of customers with different densities.

Tables 2 and 3 report the results obtained for the MPM instantiations MUr-5c1n2m-Sweep-Sweep and Ur-2c1n1m-Sweep-JSprit without and with exploration phase, respectively. Due to the large size of the instances considered, we chose Sweep for the routing of the exploration phase, since it is a simple and fast routing algorithm, although not very efficient. For this reason, it is not the purpose of the experiments presented here to compare the quality of the solutions obtained of these MPM instantiations. The algorithms of the others phases and the misplaced criteria used for the MPM instantiations were those for which we obtained the best results in the experiments performed. Columns 1 to 6 provide the information about the instances: name, number of total nodes, number of depots, total depot capacity, vehicles capacity and total customer demand, respectively. Columns 7 to 10 show the costs and the total running times (in seconds) of all phases of the MPM methodology, without and with exploration phase, respectively. The last two columns report the percentage of gap for the costs and the time ratio (the ratio between the running times observed with and without exploration).

From Tables 2 and 3 we can appreciate that the exploration phase results in a performance improvement that may depend on the routing algorithms used for the exploration and final phases. In the case of the same algorithm (MUr-5c1n2m-Sweep-Sweep), the inclusion of the exploration phase results in a better final solution for all the instances, with an average improvement of 7.18%. Although the running times increased on average 17 times, they can still be considered very good taking into account the size of the instances. In the case of different routing algorithms (Ur-2c1n1m-Sweep-JSprit), we note from Table 3

that in most instances (7 of 10) the inclusion of the exploration phase results in
a better final solution, with an improvement in costs from 0.18% to 6.83%. In
addition, empirically it seems that the inclusion of the exploration phase does
not cause a significant increase in the execution times. Indeed, in almost half of
the instances there is a marked decrease in them. This may be due to the fact
that the reassignment of customers to depots of the exploration phase simpli-
fies the final routing, i.e., less effort is needed to obtain a good quality routing.
Again, it can be seen that it is enough to use a simple and fast algorithm for the
exploration phase, and a good and eventually time consuming routing algorithm
for the final phase. However, in some cases using different routing algorithms
for the two phases can result in a higher cost final solution, as it can be seen in
Table 3.

Table 2. MUr-5c1n2m-Sweep-Sweep performance without and with exploration phase.

Inst.	Nodes	Dep.	D.Cap.	V.Cap.	Dem.	Without exp.		With exp.		% Gap	Time
						Cost	Time	Cost	Time	Cost	Ratio
L01	200	5	4250	80	3885	626409.84	0.319	591986.94	1.811	−5.50	5.68
L02	200	5	4250	80	3885	612327.72	0.043	564351.64	1.904	−7.84	44.28
L03	200	8	4400	80	3885	693744.67	0.078	690438.63	0.583	−0.48	7.47
L04	262	13	15920	500	12106	8438.75	0.060	7373.04	1.840	−12.63	30.67
L05	500	6	15000	300	12488	404082.94	0.327	364345.57	4.123	−9.83	12.61
L06	500	6	15000	300	11750	401862.56	0.440	368230.51	15.161	−8.37	34.46
L07	800	7	24500	300	22890	572609.36	0.984	546426.73	11.102	−4.57	11.28
L08	800	7	24500	300	24007	585369.28	1.978	542194.85	22.870	−7.38	11.56
L09	1050	50	50073	500	40801	482949.62	3.818	443298.2	16.027	−8.21	4.20
L10	1050	50	44450	500	40411	452780.95	1.934	420944.92	17.881	−7.03	9.25
Average										−7.18	17.15

Table 3. Ur-2c1n1m-Sweep-JSprit performance without and with exploration phase.

Inst.	Nodes	Dep.	D.Cap.	V.Cap.	Dem.	Without exp.		With exp.		% Gap	Time
						Cost	Time	Cost	Time	Cost	Ratio
L01	200	5	4250	80	3885	481912.19	10.202	487401.81	8.475	1.14	0.83
L02	200	5	4250	80	3885	468610.92	7.409	467768.58	6.902	−0.18	0.93
L03	200	8	4400	80	3885	578998.86	5.098	578998.86	4.814	0.00	0.94
L04	262	13	15920	500	12106	7022.09	23.496	6542.62	26.43	−6.83	1.12
L05	500	6	15000	300	12488	299702.93	60.413	298152.39	60.57	−0.52	1.00
L06	500	6	15000	300	11750	308062.74	56.938	312336.36	50.678	1.39	0.89
L07	800	7	24500	300	22890	466505.25	166.581	463011.41	157.925	−0.75	0.95
L08	800	7	24500	300	24007	469352.3	123.842	467293.69	125.404	−0.44	1.01
L09	1050	50	50073	500	40801	390585.03	19.974	388445.21	21.17	−0.55	1.06
L10	1050	50	44450	500	40411	414786.54	17.403	409329.19	18.793	−1.32	1.08
Average										−0.80	0.98

5   Conclusions

The Capacitated Multi-Depot Vehicle Routing Problem (CMDVRP) is an extension of the MDVRP that considers limited supply on the depots. As far as we know the CMDVRP problem has received much less attention in the literature than other MDVRP extensions. In order to solve this NP-hard problem, we introduce a Multi-Phase Methodology (MPM) that extends the well-known approach of "cluster first, route second". The most relevant feature of MPM is an intermediate exploration phase for detecting and reassigning misplaced customers based on VRP algorithms. As the VRP is a well-known and widely studied problem, the strength of the proposed MPM is to give a straightforward an efficient general framework for the direct use of VRP algorithms, in many cases publicly available and free, to solve the CMDVRP. Each MPM phase offers the possibility of choosing different algorithms depending on the specific characteristics of the problem, the problem instances, hardware limitations, time restrictions, etc. A specific selection of algorithms, for each phase, produces a particular MPM instantiation that yields a heuristic procedure to solve the CMDVRP.

From the results obtained of the numerical experiments carried out, we can conclude that the multi-phase methodology suggested can result in competitive heuristics compared to exact methods. In particular, it may be useful for users who often need to find solutions of quality in a reasonable computational time. We point out that it would be enough to use a simple and fast routing algorithm for the exploration phase, and a good and eventually time consuming routing algorithm for the final phase. We also note that in general the exploration phase produces better solutions without causing a significant increase in the execution times but, in many cases, there is a decrease in them. This may be due to the fact that the reassignment of customers to depots during the exploration phase makes that less effort is needed to obtain a good quality final routing.

The proposed multi-phase methodology enables and facilitates the use of different combinations of algorithms and the possibility to define the criterion for misplaced customers that may include geographical information, supply capacity constraints, time windows and others constraints. Therefore, it has a great potential to be adapted to specific MDVRP variants such as Periodic-VRP (PVRP), MDVRPTW or CMDVRPTW. The exploration phase of MPM allows the introduction of randomness for example in the order in which the misplaced customers are considered to be reassigned or in the reassignment strategy. We believe that the methodology could benefit from employing a randomized strategy in order to explore the solution space more extensively and eventually escape from local optimal solutions.

A possible and interesting direction for future research is to compare the proposed methodology against different solution procedures of the literature for related problems, such as MDVRP (the problem without capacity constraints on the depots). In order to make this comparison, we adapted the instances suggested by [7] and available at http://neumann.hec.ca/chairedistributique/data/mdvrp/. We consider those MDVRP instances of [7] without supply capacities on the depots nor time constraints on the routes duration, but do have restrictions on the vehicle fleet size. Preliminary results obtained in tests comparing different instances of MPM and the multiphase SFLA-PLEONS algorithm of [13], which as far as we know is one of the faster and more accurate algorithms in the literature for MDVRP, shown that MPM methodology is competitive with fastest running times. The objective is to continue doing more tests varying the instantiations of MPM methodology and also look for other instances of MDVRP in the literature.

Finally, some of the results of the numerical experiment reported, deserve a further analysis. One of them is to analyze the causes of why the addition of the exploration phase does not increase the execution times of the overall methodology, as we empirically observed in the numerical experiments reported in Tables 2 and 3. Another issue is in which cases and why it is sufficient to use a simple and fast algorithm for the exploration phase, and a good and eventually time consuming routing algorithm for the final phase, to obtain good quality solutions.

References

1. Allahyari, S., Salari, M., Vigo, D.: A hybrid metaheuristic algorithm for the multi-depot covering tour vehicle routing problem. Eur. J. Oper. Res. **242**(3), 756–768 (2015)
2. Aurenhammer, F.: Voronoi diagrams - survey of a fundamental geometric data structure. ACM Comput. Surv. **23**(3), 345–405 (1991)
3. Bektas, T., Gouveia, L.: Requiem for the Miller-Tucker-Zemlin subtour elimination constraints? Eur. J. Oper. Res. **236**(3), 820–832 (2014)
4. Calvet, L., Ferrer, A., Gomes, M.I., Juan, A.A., Masip, D.: Combining statistical learning with metaheuristics for the multi-depot vehicle routing problem with market segmentation. Comput. Ind. Eng. **94**, 93–104 (2016)
5. Clarke, G., Wright, J.W.: Scheduling of vehicles from a central depot to a number of delivery points. Oper. Res. **12**(4), 568–581 (1964)
6. Contardo, C., Martinelli, R.: A new exact algorithm for the multi-depot vehicle routing problem under capacity and route length constraints. Discret. Optim. **12**, 129–146 (2014)
7. Cordeau, J.F., Laporte, G., Mercier, A.: Improved tabu search algorithm for the handling of route duration constraints in vehicle routing problems with time windows. J. Oper. Res. Soc. **55**(5), 542–546 (2004)
8. Garey, M.R., Johnson, D.S.: Computers and Intractability: A Guide to the Theory of NP-Completeness, 1st edn. W.H. Freeman & Co., New York (1979)
9. Giosa, D., Tansini, L., Viera, O.: Assignment algorithms for the multi-depot vehicle routing problem. In: Proceedings of the 28th Conference of the Sociedad Argentina de Informática e Investigación Operativa (1999)
10. Giosa, D., Tansini, L., Viera, O.: New assignment algorithms for the multi-depot vehicle routing problem. J. Oper. Res. Soc. **53**(9), 977–984 (2002)
11. Laporte, G.: The vehicle routing problem: an overview of exact and approximate algorithms. Eur. J. Oper. Res. **59**(3), 345–358 (1992)
12. Laporte, G.: What you should know about the vehicle routing problem. Nav. Res. Logist. **54**(8), 811–819 (2007)

13. Luo, J., Chen, M.-R.: Improved shuffled frog leaping algorithm and its multi-phase model for multi-depot vehicle routing problem. Expert Syst. Appl. **41**(5), 2535–2545 (2014)
14. Montoya-Torres, J.R., Franco, J.L., Isaza, S.N., Jiménez, H.F., Herazo-Padilla, N.: A literature review on the vehicle routing problem with multiple depots. Comput. Ind. Eng. **79**, 115–129 (2015)
15. Pessoa, A., Sadykov, R., Uchoa, E., Vanderbeck, F.: A generic exact solver for vehicle routing and related problems. Math. Program. **183**, 483–523 (2020)
16. Schrimpf, G., Schneider, J., Stamm-Wilbrandt, H., Dueck, G.: Record breaking optimization results using the ruin and recreate principle. J. Comput. Phys. **159**(2), 139–171 (2000)
17. Tansini, L., Urquhart, M., Viera, O.: Comparing assignment algorithms for the multi-depot VRP. Technical report, INCO, FING, UDELAR (2001)
18. Urquhart, M., Viera, O.: A vehicle routing system supporting milk collections. Opsearch **39**, 46–54 (2002)
19. Vidal, T., Crainic, T.G., Gendreau, M., Lahrichi, N., Rei, W.: A hybrid genetic algorithm for multidepot and periodic vehicle routing problems. Oper. Res. **60**(3), 611–624 (2012)
20. Zhang, B., Li, H., Li, S., Peng, J.: Sustainable multi-depot emergency facilities location-routing problem with uncertain information. Appl. Math. Comput. **333**, 506–520 (2018)
21. Zhang, W., Chen, Z., Zhang, S., Wang, W., Yang, S., Cai, Y.: Composite multi-objective optimization on a new collaborative vehicle routing problem with shared carriers and depots. J. Clean. Prod. **274**, 1–18 (2020)
22. Zhou, L., Baldacci, R., Vigo, D., Wang, X.: A multi-depot two-echelon vehicle routing problem with delivery options arising in the last mile distribution. Eur. J. Oper. Res. **265**(2), 765–778 (2018)

A New Strategy for Collective Energy Self-consumption in the Eco-Industrial Park: Mathematical Modeling and Economic Evaluation

Hamza Gribiss$^{(\boxtimes)}$, MohammadMohsen Aghelinejad , and Farouk Yalaoui

UTT, LIST3N, Troyes, France
{hamza.gribiss,mohsen.aghelinejad,farouk.yalaoui}@utt.fr

Abstract. Renewable energies are increasingly used around the world to replace fossil energy resources such as gas, coal, and oil sources in order to reduce greenhouse gases. Eco-industrial parks promote the use and sharing of renewable energy sources between factories in a collective self-consumption framework. This article presents a new strategy of photovoltaic self-consumption in an eco-industrial park, that combines collective and individual self-consumption. This strategy has been compared with the classical configuration of self-consumption, in which factories do not share a common photovoltaic installation. Two mathematical models have been proposed and solved for these two configurations, the results show that the new strategy is more efficient than the classical configuration of individual self-consumption.

Keywords: Eco-industrial park · Renewable energy · Collective self-consumption · Mathematical modeling

1 Introduction

Energy production is mainly based on fossil energy resources such as gas, coal, and oil sources. According to the UN (United Nations Organization), the use of these resources results in global warming of $1.5\,°C$ due to the emission of greenhouse gases [1]. Among the targets set out in the European Union's (EU) climate and energy framework for 2030, is to reduce greenhouse gases emissions and to increase the share of renewable energy [2].

Energy self-consumption is an important option, which drives to increase renewable energy sources as a result of high energy prices and the emission of greenhouse gases. There are two types of self-consumption:

- The individual self-consumption, that is part of the total energy production consumed by the system. It refers to the process by which a producer consumes its energy production [3].
- The collective self-consumption, that is the case where several consumers share the same energy production. For instance, an industrial park contains

B. Dorronsoro et al. (Eds.): META 2021, CCIS 1541, pp. 212–225, 2022.
https://doi.org/10.1007/978-3-030-94216-8_16

several factories that share a photovoltaic production [4]. There is another form of collective self-consumption, particularly in eco-industrial parks. Factories exchange the surplus of energy between themselves, in terms of energy symbioses.

Eco-industrial parks (EIPs) are characterized as a set of factories located in the same geographic area, with the goal of fostering cooperation and resource sharing [5]. EIPs aim to efficiently exchange natural resources, reduce overall environmental impact, and increase economic benefits to participants [6].

This article presents a study, which combines individual and collective self-consumption in an eco-industrial park. In order to minimize energy costs and greenhouse gas emissions. The rest of this article is organized as follows. Section 2 presents the industrial symbioses involving renewable energy. Section 3 presents the problem description and Sect. 4 introduces the mathematical model. Section 5 provides the data generation for testing the model. Section 6 includes a discussion of the results. Section 7 draws conclusions with some directions for future research.

2   Related Work

The main objective of eco-industrial parks is to facilitate industrial symbiosis between a set of production units that can generate exchanges of waste, materials, and energy. Industrial symbiosis has been defined by Chertow et al. [7] as a collective commitment including physical exchanges of materials, energy, and products between factories that have geographical proximity. In this context, Butturi et al. [8] proposed a multi-objective optimization to evaluate energy symbiosis, that includes the integration of renewable energy sources within an eco-industrial park and considering both economic and environmental issues. They discuss three scenarios that provide individual company and park managers with relevant information, which support the decision-making regarding the economic sustainability and environmental impacts of energy symbiosis. Jiang et al. [9] presented a genetic algorithm to solve a multi-objective optimization, that propose an exchange the electricity between four parks in absence of grid power. Their aim was to minimize a power interruption, storage system cost, and customer dissatisfaction. Heendeniya [10] proposed an agent-based model to exchange energy between prosumers, which have their own PV power generation and a battery storage. Each agent tries individually and collectively to maximize self-consumption of renewable energy.

The economic feasibility of self-consumption in eco-industrial parks and remote areas has been studied in several papers. Among these works, Contreras et al. [11] present a cooperative planning framework that integrates long-term planning and short-term operation of an energy collective composed of consumers sharing a photovoltaic and storage system. Their objective was to determine the optimal size of the PV plus storage system, that reduces total costs. Long-term planning gives each consumer an idea of how much they can save, which allows them to decide to join the collective or not. Pedrero et al. [12] presented an economic evaluation of shared self-consumption of PV installations between three halls with the addition of the option to sell surplus energy. They prove that the economic feasibility depends largely on the compensation for the electricity fed into the grid.

A study on the integration of renewable energy in eco-industrial parks in the literature has been treated by Butturi et al. [13]. The result of this study shows that a few articles have considered the integration of renewable energy. Among the works that have been published after this research [13], we find Jiang et al. [9], which discuss in their paper the exchange of renewable energy between four parks with the integration of a storage system. Butturi et al. [8] treat the case of renewable energy exchange between factories. In another study, Pedredro et al. [12] deal with the case of collective self-consumption with the option of selling the surplus energy between 3 factories that share a photovoltaic installation.

The result of this state of the art shows that there is a lack of articles that deal with the combination of individual and collective self-consumption within an eco-industrial park, as well as few papers have considered in the same study the storage option and the option of selling the surplus energy.

In this paper, a new strategy of energy self-consumption. in eco-industrial park is introduced. It allows the merge of both individual and collective self-consumption with the integration of storage systems and the addition of the option of selling surplus energy.

3   Problem Definition

In this section, the eco-industrial park's strategy and the classic individual self-consumption configuration are presented. The structure of the strategy of the eco-industrial park is as follows:

- Each factory has its own photovoltaic production that provides energy to satisfy the demands. Excess energy is either stored in the factory's battery or sold to the grid.
- In case the factory's self-production is not sufficient to guarantee its energy requirements, the factory relies on the shared photovoltaic production or the shared battery.
- The common photovoltaic production and the common battery provide a percentage of their energy to each factory. This percentage depends on the investment cost of each factory for the creation of the park. The surplus energy from the common production is either stored in the common battery or sold to the grid.

– The appeal to the main grid is made in case of an emergency where the factory's own production, its battery, the common production, and the common battery are not enough to guarantee the energy demands needs in the factories.

The Fig. 1 represents the schema of this strategy for a case of three factories.

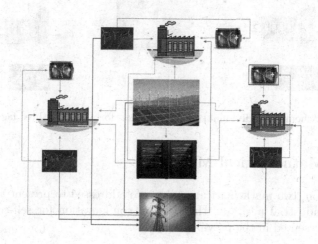

Fig. 1. Schema of strategy for a case of three factories.

The objective is to compare this strategy with the classical configuration (individual self-consumption) which is already studied in several articles in the literature. This configuration is very present in residential buildings that use photovoltaic panels with the integration of batteries [3]. As an example to this study, Braun et al. [14] used a lithium-ion battery to increase self-consumed photovoltaic energy with the addition of the option of selling the surplus energy.

The structure of this configuration in our study is as follows:

– Each factory has its own photovoltaic production that provides energy to satisfy the demands. Excess energy is either stored in the factory's battery or sold to the grid.
– In case the factory's self-production is not sufficient to guarantee its energy requirements, the factory draws energy from the grid.

The Fig. 2 represents the schema of classical individual self-consumption for a case of three factories.

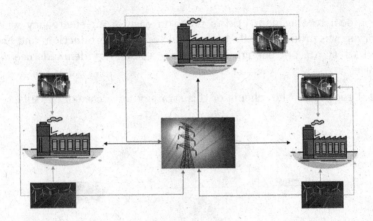

Fig. 2. Schema of individual self-consumption for a case of three factories.

4   The Mathematical Models

In this section, two mathematical models are addressed to present the strategy of the eco-industrial park and the classical configuration for self-consumption which are presented in the previous section.

4.1   Parameters and Decision Variables

Sets:

$j = \{1, \ldots, J\}$: set of factories in the eco-industrial park

$t = \{1, \ldots, H\}$: set of the time period (in hours)

$i = \{1, \ldots, I\}$: set of PV installation in the park

Parameters:

$D_{j,t}[KWh]$: Energy demand of factory j at period t

$Q_{i,t}^{p}[KWh]$: Amount of photovoltaic energy available in the shared source i at period t

$Q_{j,t}^{f}[KWh]$: Amount of photovoltaic energy available in the factory j at period t

$P_{t}^{g}[\text{€}/KWh]$: Price of energy from the grid at period t

$P_{t}^{s}[\text{€}/KWh]$: Price of energy sold to the grid at period t

$P_{t}^{f}[\text{€}/KWh]$: Price of energy drawn from the factory's production at period t

$P_{t}^{p}[\text{€}/KWh]$: Price of energy drawn from the shared production at period t

$P_{t}^{bf}[\text{€}/KWh]$: Price of energy drawn from the factory's battery at period t

$P_{t}^{bp}[\text{€}/KWh]$: Price of energy drawn from the shared battery at period t

$SOC_{j}^{max}[kWh]$: Maximum state of charge of the factory's battery j

$SOC_{j}^{min}[kWh]$: Minimum state of charge of the factory's battery j

$SOCP^{max}[kWh]$: Maximum state of charge of the shared battery

$SOCP^{min}[kWh]$: Minimum state of charge of the shared battery

η^{char}: Losses due to the battery's charging

η^{dech}: Losses due to the battery's discharge

$IC_j^f[\text{€}]$: Investment costs related to the factory j for its own photovoltaic energy and battery during the horizon H

$IC_j^p[\text{€}]$: Investment costs related to the factory j for the common photovoltaic production and battery during the horizon H

$IC^p[\text{€}]$: Total investment costs of the park during the horizon H

$Pr_j^f[\%]$: The contribution rate of the factory j to the construction of the park

Decision variables:

$E_{j,t}^f[KWh]$: Amount of energy which is drawn by factory j at period t from its PV production

$E_{i,j,t}^p[KWh]$: Amount of energy which is drawn by factory j at period t from the common source i

$E_{j,t}^g[KWh]$: Amount of energy which is drawn by factory j at period t from grid

$E_{j,t}^{dbf}[KWh]$: Amount of energy which is drawn by factory j at period t from its battery

$E_{j,t}^{dbp}[KWh]$: Amount of energy which is drawn by factory j at period t from the common battery

$E_{j,t}^{cbf}[KWh]$: Amount of produced energy by the factory j at period t stored in its battery

$E_{i,t}^{cbp}[KWh]$: Amount of produced energy by the common source i at period t stored in the common battery

$E_{j,t}^{sf}[KWh]$: Amount of produced energy by the factory j at period t that is sold to the grid

$E_{t,i}^{sp}[KWh]$: Amount of produced energy by the common source i at period t that is sold to the grid

$E_t^{sp}[KWh]$: Amount of produced energy in park at period t that is sold to the grid

SOC_{jt}: State of charge of the factory's battery j at period t

$SOCP_t$: State of charge of the common battery j at period t

$c_t^p: = 1$ if the common battery is charging at period t. 0 otherwise

$d_t^p: = 1$ if the common battery is discharging at period t. 0 otherwise

$c_{j,t}^f: = 1$ if the factory's battery j is charging at period t. 0 otherwise

$d_{j,t}^f: = 1$ if the factory's battery j is discharging at period t. 0 otherwise

In the following, the model of the individual self-consumption is presented as model 1 and the model of the individual and collective self-consumption is presented as model 2.

4.2   Objective Function of Model 1

The objective function aims to minimize the energy cost of the factories. It is calculated by subtracting the following two blocks:

– The first is the summation of the variable costs of energy which are: the costs of purchasing energy from the grid, the factories productions, and the factories batteries.
– The second is the sale of the surplus energy to the grid by the factories.

$$\min Z_1 = \sum_{j=1}^{J} \sum_{t=1}^{H} (P_t^g \times E_{j,t}^g + P_t^f \times E_{j,t}^f + P_t^{bf} \times E_{j,t}^{dbf} - P_t^s \times E_{j,t}^{sf}) \quad (1)$$

4.3   Constraints of Model 1

Constraint (2) ensures that the total demand of each factory j is satisfied by the energy sources available in factory j and the grid at period t.

$$E_{j,t}^f + E_{j,t}^{dbf} + E_{j,t}^g = D_{j,t} \quad \forall j, t \quad (2)$$

Constraint (3) ensures that the sum of the quantities of energy drawn by each factory j, stored in the factory's battery j, and sold to the grid must be equal to the quantity of energy available in the factory's production j at period t.

$$E_{j,t}^f + E_{j,t}^{cbf} + E_{j,t}^{sf} = Q_{j,t}^f \quad \forall j, t \quad (3)$$

The constraints (4), (5), and (6) represent the state of charge initial, final, and at period t respectively for the factory's battery j.

$$SOC_{j,0} = SOC_j^{min} \quad (4)$$

$$SOC_{j,T} = SOC_j^{min} \quad (5)$$

$$SOC_{j,t} = SOC_{j,(t-1)} + \eta^{char} \times E_{j,t}^{cbf} - 1/\eta^{dech} \times E_{j,t}^{dbf} \quad \forall j, t \quad (6)$$

Constraint (7) ensures that the factory's battery j is protected against accelerated aging.

$$SOC_j^{min} \leq SOC_{j,t} \leq SOC_j^{max} \quad \forall j, t \quad (7)$$

Additionally, the amount of charging and discharging of the factories batteries must also meet the upper and lower bound constraints. Constraints (8) and (9) refer to the maximum to be charged in the factory's battery j at period t and constraints (10) and (11) represent the maximum to be discharged in the factory's battery j at period t.

$$E_{j,t}^{cbf} \leq SOC_j^{max} \times c_{j,t}^f \quad \forall j, t \quad (8)$$

$$E_{j,t}^{cbf} \geq c_{j,t}^f \quad \forall j, t \quad (9)$$

$$E_{j,t}^{dbf} \leq SOC_j^{max} \times d_{j,t}^f \quad \forall j, t \quad (10)$$

$$E_{j,t}^{dbf} \geq d_{j,t}^f \quad \forall j, t \quad (11)$$

Constraint (12) ensures the choice between charging or discharging of the factory's battery j at period t.

$$c_{j,t}^f + d_{j,t}^f \leq 1 \quad \forall j, t \quad (12)$$

4.4   Objective Function of Model 2

The aim of the objective function of model 2 is the same as of model 1, it minimizes the energy cost of the factories in the park. Equation (13) represents this objective function.

$$\min Z_2 = \sum_{j=1}^{J} \sum_{t=1}^{H} (P_t^g \times E_{j,t}^g + + P_t^f \times E_{j,t}^f + P_t^{bf} \times E_{j,t}^{dbf} + P_t^p \times \sum_{i=1}^{I}(E_{i,j,t}^p)$$
$$+ P_t^{bp} \times E_{j,t}^{dbp} - P_t^s \times (E_{j,t}^{sf} + E_t^{sp})) \tag{13}$$

4.5   Constraints of Model 2

Constraints (3–12) of model 1 are applied to model 2. In addition, the following constraints have been exclusively applied to model 2.

Constraint (14) ensures that the total demand of each factory j is satisfied by the energy sources available in the park, the factory j, and the grid at period t.

$$E_{j,t}^f + E_{j,t}^{dbf} + \sum_{i=1}^{I}(E_{i,j,t}^p) + E_{j,t}^{dbp} + E_{j,t}^g = D_{j,t} \quad \forall j, t \tag{14}$$

Constraint (15) ensures that the sum of the quantities of energy demands of all factories is satisfied by the common source i, the energy stored in the common battery from source i, and the energy sold to the grid from source i must be equal to the quantity of energy available in the shared source i at period t.

$$\sum_{j=1}^{J}(E_{i,j,t}^p) + E_{i,t}^{cbp} + E_{t,i}^{sp} = Q_{i,t}^p \quad \forall t, i \tag{15}$$

Constraint (16) represents the total energy sold to the grid by the shared production.

$$\sum_{i=1}^{I}(E_{t,i}^{sp}) = E_t^{sp} \quad \forall t \tag{16}$$

Constraint (17) represents the percentage of energy to be drawn from the common production and the common battery by each factory j during the horizon H.

$$\sum_{t=1}^{H}(E_{j,t}^{dbp} + \sum_{i=1}^{I}(E_{i,j,t}^p)) = Pr_j^f \times \sum_{t=1}^{H}\sum_{i=1}^{I}(Q_{i,t}^p) \quad \forall j \tag{17}$$

The constraints (18), (19), and (20) represent the state of charge initial, final, and at period t respectively for the shared battery.

$$SOCP_0 = SOCP^{min} \tag{18}$$

$$SOCP_T = SOCP^{min} \tag{19}$$

$$SOCP_t = SOCP_{(t-1)} + \eta^{char} \times \sum_{i=1}^{I}(E_{i,t}^{cbp})$$

$$- 1/\eta^{dech} \times \sum_{j=1}^{J}(E_{j,t}^{dbp}) \quad \forall t \qquad (20)$$

Constraint (21) ensures that the shared battery is protected against accelerated aging.

$$SOCP^{min} \leq SOCP_t \leq SOCP^{max} \quad \forall t \qquad (21)$$

Additionally, the amount of charging and discharging of the common battery must also meet the constraints of the upper and lower limits. Constraints (22) and (23) refer to the maximum to be charged in the common battery at period t and Constraints (24) and (25) represent the maximum to be discharged in the common battery at period t.

$$\sum_{i=1}^{I}(E_{i,t}^{cbp}) \leq SOCP^{max} \times c_t^p \quad \forall t \qquad (22)$$

$$\sum_{i=1}^{I}(E_{i,t}^{cbp}) \geq c_t^p \quad \forall t \qquad (23)$$

$$\sum_{j=1}^{J}(E_{j,t}^{dbp}) \geq d_t^p \quad \forall t \qquad (24)$$

$$\sum_{j=1}^{J}(E_{j,t}^{dbp}) \geq SOCP^{max} \times d_t^p \quad \forall t \qquad (25)$$

Constraint (26) ensures the choice between charging or discharging of the shared battery at period t.

$$c_t^p + d_t^p \leq 1 \quad \forall t \qquad (26)$$

5   Data Generation

This section presents how the model data was generated, such as photovoltaic installation, battery size, and investment costs.

5.1   Photovoltaic Installation in Factories

In this study, the maximum size of photovoltaic production will be installed on the roofs of factories is considered. As a hypothesis, the available surface on the roofs of these factories taken into account to install photovoltaic panels is $S^{min} \leq S \leq S^{max}$ where $S^{min} = 800\text{m}^2$ and $S^{max} = 1200\text{m}^2$ Based on [15] a 1.9 m^2 monocristallin solar panel can produce 365 W, so the maximum size of the photovoltaic installation that can be placed in a surface S is $\alpha = \frac{S \times 365}{1,9}$.

To calculate the amount of energy $Q_{j,t}^f$ produced with a PV installation α, in each period t during the horizon H, the European Commission's Photovoltaic Geographic Information System (PVGIS) [16] is used. The PVGIS-SARAH radiation database is chosen, it can offer PV load profiles with a resolution of one hour between 2005 and 2016, which in turn were used to generate an average annual PV load profile for each factory roof.

To summarize, for each factory j of surface S_j, it is possible to install a PV production size α_j that gives an amount of energy $Q_{j,t}^f$ at each period t.

5.2 Photovoltaic Installation in the Park

The size of the PV installation in the park is chosen with the following method:

- Calculating the difference between the total demand and the total energy produced by all factories in one year, which is defined by $R = \sum_{t=1}^{8760} \sum_{j=1}^{J} D_{j,t} - \sum_{t=1}^{8760} \sum_{j=1}^{J} Q_{j,t}^f$
- Using PVGIS, the determination of the size of the PV installation that produces a percentage k of R, i.e.: $\sum_{t=1}^{8760} \sum_{i=1}^{I} Q_{i,t}^p = k \times R$. This size is used to calculate the amount of energy produced in the park over 4 horizons (1 month, 1 season, 2 seasons, and 4 seasons).

In this study, the cases where k is 20% 40%, 60%, and 80% are compared.

5.3 Battery Size in the Factories and in the Park

In a study of PV self-consumption by Luthander et al. [3], they report that PV self-consumption can be increased by 13–24% by using a battery capacity of 0.5–1 kWh for each KW of PV power installed. In this case study, a 1 kWh lithium battery is installed for every 1 kW of PV power installed.

5.4 Investment Cost of the Factories

For each factory, there are two types of investment costs:

- Fixed investment costs for its own photovoltaic energy and battery during the horizon H.
- Fixed investment costs for the shared photovoltaic production and battery during the horizon H.

To calculate the investment costs in the factories and in the park, the data of Pedrero et al. is used [12], represented in Tables 1 and 2.

Table 1. PV installations reference costs

PV power	Reference cost [€/W]
≤ 10 kW	1.5
10 kW−100 kW	0.9
100 kW−1 MW	0.75

Table 2. Economic parameters for the calculation of investment costs

Parameter	Value
PV modules service life	25 (year)
Inverter service life	15 (year)
Inverter cost	0.2€/W
Maintenance cost	0.02€/(W × year)

To calculate the contribution cost for each factory, Table 3 is relied upon.

Table 3. Percentage of contribution for each factory

	Total demand	Percentage of contribution
Factory j	$\sum_{t=1}^{H} D_{j,t}$	$Pr_j^f = \frac{\sum_{t=1}^{H} D_{j,t}}{\sum_{t=1}^{H} \sum_{j=1}^{J} D_{j,t}}$

5.5 Different Energy Costs

The variation in electricity prices over the optimisation horizon can affect the total energy cost. In this paper, time-of-use (TOU) pricing is put to use to balance electricity supply and demand. The purchase price of the electricity grid is the most expensive and the price of the energy from the factory is the cheapest. The prices are classified in this order:

$$P_t^f \leq P_t^{bf} \leq P_t^p \leq P_t^{bp} \leq P_t^g \ \ \forall t$$

6 Numerical Study

In this section, illustrative examples are considered to validate and evaluate the presented models, which are solved by CPLEX on an Intel Core i5 with 2.7 GHz and 8 GB RAM.

To compare the results between the proposed new strategy and the classical configuration of individual self-consumption, the ratio between the investment and the price paid by all factories at the end of the horizon H is calculated. $X = \frac{IC1}{G1}$ and $Y = \frac{IC2}{G2}$ represent this ratio in the case of individual self-consumption and the strategy respectively with:

- $IC1$: Total investment cost of the factories during the horizon H for the case of individual self-consumption.
- $G1$: Total cost paid by the factories during the horizon H for the case of the individual self-consumption.
- $IC2$: Total investment cost of the factories during the horizon H for the case of the strategy.
- $G2$: Total cost paid by the factories during the horizon H for the case of the strategy.

In the rest of this paper, four different cases of the eco-industrial park are used. The first one contains 3 factories, the second 6 factories, the third 9 factories, and the fourth 15 factories. Each case is tested over 4 horizons (1 month, 1 season, 2 seasons, and 4 seasons) which are presented in hours. The values represented in the following tables are the average value of the gap between X and Y over the 4 horizons.

The gap can be calculated by $\frac{Y-X}{X} \times 100$. For example, the value 235.4 presented in Table 4 represents this gap in the case of 3 factories and $k = 40\%$. It was calculated using the previous gap formula with $X = 0,1184$ and $Y = 0,39712$. The more this gap is greater, the considered strategy is more performed.

Variation of the Size of the Photovoltaic Installation in the Park

Table 4 represents the average values of the gap during the four horizons for each case of the eco-industrial park by varying the percentage k of the photovoltaic installation of the park. As a result, it can be concluded that:

- In each instance, the use of the proposed strategy model is better than the individual self-consumption model.
- Despite the increase in the number of factories, there is a small decrease of the gap, which gives the possibility to add several factories in the same park without having the problem of decreasing the gap.
- By increasing the energy of the park by 20% the gap increases between 52,44% and 123,14%.

Table 4. Variation of the size of the photovoltaic installation in the park

	3 factories (%)	6 factories (%)	9 factories (%)	15 factories (%)	Average (%)
$k = 20\%$	105,50	105,51	102,60	102,59	104,05
$k = 40\%$	235,40	235,44	228,65	228,62	232,03
$k = 60\%$	398,19	386,38	386,16	386,08	389,20
$k = 80\%$	608,16	603,25	588,69	588,58	597,17
Average	336,81	332,64	326,52	326,47	

Variation in the Selling Price of Surplus Energy

Due to low feed-in tariffs [10] especially for large installations, three categories of energy selling price are considered such as: $P_t^s = 1/10 \times P_t^g$, $P_t^s = 1/5 \times P_t^g$ and $P_t^s = P_t^g$ where P_t^s: Price of energy sold to the grid at period t and P_t^g: Price of energy from the grid at period t.

Table 5 represents the average values of the gap during the four horizons for each case of the eco-industrial park by varying the selling price of the surplus energy.

Table 5. Variation in the selling price of surplus energy

	3 factories (%)	6 factories (%)	9 factories (%)	15 factories (%)	Average
$P_t^s = 1/10 \times P_t^g$	231	231,04	224,28	224,25	227,64
$P_t^s = 1/5 \times P_t^g$	235,40	235,44	228,65	228,62	232,02
$P_t^s = P_t^g$	298,99	299,00	291,13	266,79	288,97
Average	255,13	255,16	248,13	239,89	

It is concluded that:

- The proposed strategy is more efficient than the individual self-consumption in each case.
- The increase in the selling price of surplus energy increases the gap between the proposed strategy and the classical configuration of individual self-consumption.

The proposed strategy gives better results than the classical configuration of individual self-consumption even in the months when there is little photovoltaic production such as January.

7   Conclusion

This paper develops two mathematical models, the first one represents a new strategy of self-consumption in eco-industrial parks and the second one defines the classical configuration of individual self-consumption. In both models, the option of storage and sale of surplus energy have been addressed. This study represents a step forward in the under-investigated field regarding the integration of renewable energy in eco-industrial park [13]. The two models were tested and compared, the results show that with this new strategy the factories can reduce the price of electricity compared to the classical configuration of individual self-consumption.

According to [17] the necessary condition to justify the creation of an eco-industrial park, is to prove that the benefits achieved by working with a collective strategy of factories are superior to the benefits achieved by working as a single factory. This study shows that investing collectively among the factories is more efficient than investing alone.

With this strategy, several factories can be in the same park because the results show that despite the increase in the number of factories, there is a small decrease in the gap and the results obtained by this new strategy are still more efficient.

References

1. Nations unies. https://www.un.org/fr/. Accessed 12 Jan 2021
2. European unions. https://europa.eu/. Accessed 24 Feb 2021
3. Luthander, R., Widén, J., Nilsson, D., Palm, J.: Photovoltaic self-consumption in buildings: a review. Appl. Energy **142**, 80–94 (2015)
4. Menniti, D., Sorrentino, N., Pinnarelli, A., Mendicino, S., Vizza, P., Polizzi, G.: A blockchain based incentive mechanism for increasing collective self-consumption in a nonsumer community. In: 2020 17th International Conference on the European Energy Market (EEM), pp. 1–6. IEEE (2020)
5. Le Tellier, M., Berrah, L., Stutz, B., Audy, J.-F., Barnabé, S.: Towards sustainable business parks: a literature review and a systemic model. J. Clean. Prod. **216**, 129–138 (2019)
6. Liu, Z.: Co-benefits accounting for the implementation of eco-industrial development strategies in the scale of industrial park based on emergy analysis. Renew. Sustain. Energy Rev. **81**, 1522–1529 (2018)
7. Chertow, M.R.: Industrial symbiosis: literature and taxonomy. Ann. Rev. Energy Environ. **25**(1), 313–337 (2000)
8. Butturi, M.A., Sellitto, M.A., Lolli, F., Balugani, E., Neri, A.: A model for renewable energy symbiosis networks in eco-industrial parks. IFAC-PapersOnLine **53**(2), 13137–13142 (2020)
9. Jiang, Y., Yang, J., Wang, C., Cao, Y.: Electricity optimal scheduling strategy considering multiple parks shared energy in the absence of grid power supply. Int. Trans. Electr. Energy Syst. **30**(12), e12634 (2020)
10. Heendeniya, C.B.: Agent-based modeling of a rule-based community energy sharing concept. In: E3S Web of Conferences, vol. 239. EDP Sciences (2021)
11. Contreras-Ocaña, J.E., Singh, A., Bésanger, Y., Wurtz, F.: Integrated planning of a solar/storage collective. IEEE Trans. Smart Grid **12**(1), 215–226 (2020)
12. Pedrero, J., Hernández, P., Martínez, Á.: Economic evaluation of PV installations for self-consumption in industrial parks. Energies **14**(3), 728 (2021)
13. Butturi, M.A., Lolli, F., Sellitto, M.A., Balugani, E., Gamberini, R., Rimini, B.: Renewable energy in eco-industrial parks and urban-industrial symbiosis: a literature review and a conceptual synthesis. Appl. Energy **255**, 113825 (2019)
14. Braun, M., Büdenbender, K., Magnor, D., Jossen, A.: Photovoltaic self-consumption in Germany: using lithium-ion storage to increase self-consumed photovoltaic energy (2009)
15. Pv surface. https://fr.electrical-installation.org/. Accessed 1 Mar 2021
16. Pvgis. https://re.jrc.ec.europa.eu/. Accessed 15 Mar 2021
17. Boix, M., Montastruc, L., Azzaro-Pantel, C., Domenech, S.: Optimization methods applied to the design of eco-industrial parks: a literature review. J. Clean. Prod. **87**, 303–317 (2015)

Author Index

Printed in the United States
by Baker & Taylor Publisher Services